OXFORD READINGS IN PH

6.95

KANT ON PURE REASON

Also published in this series

Other volumes are in preparation

KANT
ON PURE REASON

Edited by
Ralph C. S. Walker

OXFORD UNIVERSITY PRESS
1982

Oxford University Press, Walton Street, Oxford OX2 6DP

London Glasgow New York Toronto
Delhi Bombay Calcutta Madras Karachi
Kuala Lumpur Singapore Hong Kong Tokyo
Nairobi Dar es Salaam Cape Town
Melbourne Auckland

and associate companies in
Beirut Berlin Ibadan Mexico City

Published in the United States by
Oxford University Press, New York

Introduction and selection © Oxford University Press 1982

British Library Cataloguing in Publication Data

Kant on pure reason. – (Oxford readings in philosophy)
  1. Kant, Immanuel. Kritik der reinen
  Vernunft – Criticism and interpretation
  I. Walker, Ralph C.S.
  121 B2779

  ISBN 0-19-875056-0

Library of Congress Cataloging in Publication Data

Kant on pure reason
  (Oxford readings in philosophy)
  Bibliography: p.
  Includes index.
  1. Kant, Immanuel, 1724–1804. Kritik der reinen Vernunft. 2. Knowledge,
Theory of. 3. Causation. 4. Reason. I. Walker, Ralph Charles
Sutherland. II. Series.
  B2779.K36   121   81-18849

  ISBN 0-19-875056-0 (pbk.) AACR2

Typeset by Oxford Verbatim Limited
Printed in the United States of America

CONTENTS

INTRODUCTION

KANT saw philosophy as faced with a fundamental problem, the resolution of which required a critique of pure reason. This problem was to discover how, and to what extent, it is possible for us to have knowledge which is both synthetic and *a priori*. Put in that form it is not a problem which most philosophers would regard as particularly pressing nowadays, yet there has never been more interest in Kant's first *Critique* than there is now. This is because questions about the nature and limits of human knowledge remain central in philosophy, and in order to answer what at first looks like a specific question about a particular kind of knowledge Kant mounts a full-scale investigation of our cognitive faculties and of the conditions under which knowledge is possible at all. It is also because of the character of the solutions that he gave, solutions which are in some ways very attractive and in other ways puzzling and obscure.

To his question about synthetic *a priori* knowledge he has really two distinct answers, both of them highly original and each by itself sufficient to earn for him that honoured place in the epistemologists' pantheon which is reserved for those who find a new reply to scepticism. One is the method of transcendental arguments; the other is the metaphysics of transcendental idealism. At different times and places since his day one or other of these has caught the imagination of philosophers, and absorbed their attention as they have sought to develop, to complete, and to make their own that aspect of his work. It is perhaps a peculiarity of our own day that both approaches are taken very seriously, not just by Kantian scholars, but by philosophers attempting to deal with the fundamental issues of metaphysics and epistemology. I shall say something briefly about each of them in turn.

Transcendental Arguments. There is no shortage of controversy over how transcendental arguments should be formulated or how much we can hope to achieve by using them. We can however say that they are arguments which start from the consideration that we have experience, or experience of some very general kind; since the aim is to make headway against the epistemological sceptic, it is important that the kind of experience in question be characterized in a way the sceptic cannot reject without rendering his own position untenable. They then proceed by showing that unless a certain thing were the case there could not be any such experience; and conclude that that thing is indeed the case. According to Kant such conclusions are synthetic and *a priori*, synthetic because they give us information about how things are and *a priori* because they are established without empirical investigation.

It is not wholly clear how we are supposed to prove that something is required for experience, or experience of a given sort. From what has just been said it is evident that empirical methods are out of the question; we must establish it *a priori*. Some philosophers, and some interpreters of Kant, have thought it must be a matter of analysing the concept of experience, and of showing the propostion 'If there is experience of such-and-such a kind, then . . .' to be analytic. Others, doubtful about the possibility of such an analysis or fearing that results achieved in this way would be bound to be trivial, have rejected this suggestion. The difficulty about their view as a matter of Kant interpretation is that they must then assign to the 'If . . . then . . .' proposition a synthetic *a priori* status, whereas Kant wants to reserve that status for the *conclusions* of transcendental arguments – though he may not have thought the matter through.[1] The difficulty about it philosophically is that it leaves it very obscure how we can know that such a propostion is true. In other words it raises again the question 'How is synthetic *a priori* knowledge possible?', which the method of transcendental arguments was meant to resolve.

But whichever view one takes about this, the greatest question facing the method of transcendental arguments is how much it can achieve. Kant's own arguments are obscure, and the one which appears to have the central place in the *Critique* – the transcendental deduction of the categories – is among the most difficult in philosophical literature. In recent years many philosophers have tried to establish worthwhile conclusions either by finding valid transcendental arguments in Kant's work or by developing new ones of their own; and here the work of P. F. Strawson has been particularly important. In his books *Individuals*[2] and *The Bounds of Sense*[3] he provided a range of transcendental arguments which were designed to confute the epistemological sceptic on the most central and troublesome issues; at the same time they were meant to constitute an exercise in descriptive metaphysics, an exploration of the most general and indispensable features of our conceptual scheme.

Strawson's arguments are Kantian in character, though some of them are more directly due to Kant than others; his conclusions are Kantian as well, though in certain cases he seeks to establish less than Kant did and in certain others more. But there is also a fundamental respect in which his conclusions are different from Kant's, for Strawson has no sympathy with transcendental idealism. The synthetic *a priori* propositions Kant thinks he can establish by his transcendental arguments are supposed to express truths about the world as we experience it, the phenomenal world. The phenomenal world is in a sense our own creation, since it is the product of the operation of our cognitive faculties upon the data we receive in sense. It is empirically real, since we can all agree about it and it meets our ordinary and scientific standards of objectivity; but viewed from a more ultimate standpoint, viewed transcendentally, it is ideal. Kant does not consider that transcendental arguments can ever establish conclusions about the underlying noumenal reality of things as they are in themselves, apparently because he cannot accept the possibility that our possession of experience should be contingent on conditions wholly independent of us. For these conditions might hold accidentally, and not as a matter of metaphysical necessity; Kant cannot bring himself to admit that our possession of experience may entirely depend on conditions which merely happen to hold. Strawson on the other hand sees no objection to admitting our dependence on conditions quite outside ourselves, and he completely rejects the distinction of phenomenal and noumenal worlds and with it all Kant's talk of the mind creating the familiar world of the empirically real. He is not alone in repudiating this side of Kant's thought; there is a long-standing tradition which seeks to reconstruct Kant without the transcendental idealism, regarding that as a sad lapse on the part of an otherwise great man. At one time, indeed, it was commonly argued that at the height of his Critical period Kant did not really subscribe to so regrettable a theory, but this has become hard to sustain since the work of Adickes.[4]

At first sight Strawson's position looks both plausible and attractive. Kant's mysterious talk of the mind making nature, and all the complexities of the resulting picture of two worlds whose interrelation is bound to be problematic, sounds at its best gratuitous and at its worst a fine example of just that kind of unbridled speculation which Kant in his best moments was anxious to sweep away. And I think that even the most dedicated Kantian must admit that Strawson is right to say there is no reason *in principle* why transcendental arguments should not establish conclusions about how the world must really be, and not just about the phenomenal world, the world of appearances. Here Kant was simply mistaken.

The victory may be less substantial than it seems, however. In the article included in this collection Barry Stroud examines certain transcendental arguments put forward by Strawson and certain others put forward by Shoemaker, and finds them lacking in an important respect. If Stroud is right, these arguments cannot reach their intended conclusions except by relying on some version of the verification principle. Without that principle the most they can show is that we must possess certain *concepts* or have certain *beliefs*. If this is correct it is important, especially since Stroud's point is meant to apply more generally than just to these specific arguments. Though he himself does not view it in this way, it can be seen as providing valuable ammunition for the defender of transcendental idealism; for it is the transcendental idealist who holds that we must possess certain concepts and beliefs, but have no way of determining whether they correspond to any features of the world as it actually is. Strawson wanted to repudiate this, but not by relying on so questionable a resource as the verification principle. If his arguments can only establish how we must believe things to be, and if the question how they really are cannot be answered except by adding an unwanted verificationist premiss, it would rather look as though something like transcendental idealism might not be so absurd after all. Stroud, admittedly, takes himself to be attacking Kant as much as anyone, for at the end of his article he describes him as making Strawson's mistake and claiming to show how the world must actually be. But Kant's conclusions are about how things are in the phenomenal world, and between how things are in that 'world' and how we must all believe them to be it is hard to find very much difference.

Stroud does not establish that it is impossible in principle for transcendental arguments to show how things must be independently of us and of our beliefs about them, but rather asks how they could manage to achieve this without relying on some form of verificationism. It may be that this can be answered, at least for a limited range of cases; and arguably Kant himself was committed, though he was rather confused about the matter, to a transcendental argument to the real existence of an active self. But it may also be (as Stroud seems to suspect) that the amount that valid transcendental arguments can actually establish about the world is extremely small. At all events it raises again, in a strong form, the question we met before: how much can transcendental arguments prove? In the last resort I think the question can only be handled by looking at the particular examples Kant and others have put forward, to see how much they do prove. But the critics have raised enough doubts about the method to encourage us to have a look at transcendental idealism, Kant's alternative solution to the problem of synthetic *a priori* knowledge.

Transcendental Idealism. The solution provided by transcendental idealism is that synthetic *a priori* knowledge is possible because we ourselves read the appropriate truths into the world: we construct the world of appearances in such a way that they hold of it. Kant thought transcendental idealism required supplementation by transcendental arguments because only by showing them to be indispensable can we distinguish the principles which are thus constitutive of the phenomenal world from purely fictional beliefs we may all happen to share. This is a very moot point, but it helps to explain why he always thought of transcendental idealism and transcendental arguments together and failed to see the possibility of using transcendental arguments to show how things must be in themselves.

The main difficulty in trying to assess transcendental idealism is that of determining what exactly the theory amounts to. This may look easy enough at first sight, but it is made difficult by Kant's habit of expressing his position in terminology which is more suggestive than precise. Just what is meant by saying that the understanding prescribes its laws to nature, or that we ourselves introduce the order and regularity which we find in things?[5] Some, like Strawson, feel that no sense can be made of such talk, and that it should have been ruled out as illegitimate by Kant's own standards. Others feel that at least a part of it, if not the whole, can be given a harmless interpretation under which it may turn out to be more or less acceptable.

Thus J. F. Bennett takes Kant to have been a fairly orthodox phenomenalist, for whom the external world is a logical construction out of the data of experience.[6] He actually equates transcendental idealism with phenomenalism, though he is of course aware that Kant allows a place to the thing in itself which is not susceptible of any phenomenalistic treatment. Like Strawson he regards this as an unfortunate aberration on Kant's part, and he does his best to 'salvage' as many such passages as possible by reinterpreting them in a way he considers more satisfactory. His essay in this collection makes use of his phenomenalist account in elucidating Kant's solution to the First Antinomy, though I think its essential point can stand even if one rejects the claim that Kant was a phenomenalist, or a phenomenalist in Bennett's sense. (I shall return to this below.)

In fact it is difficult to deny that Kant's treatment of the world of appearances is in some sense phenomenalistic, since he regards it as a construction which our minds effect by working on the data given in sensory intuition. But it is arguable that to ascribe to him a traditional phenomenalist view of the matter is to do little justice to the subtlety and complexity of his approach. The traditional phenomenalist regards statements about the objects we can perceive as logically equivalent to statements about the actual or potential data of experience; on the right hand side of the equivalence we ought to have nothing but descriptions of the content of actual and possible experiences, together with operators of a purely logical kind. For Kant, on the other hand, the construction must be carried out by means of the categories, which despite their close relation to the logical forms of judgement do not appear to be just logical operators. The traditional view also leaves no place for the idea of the transcendental object, which at least in the first edition of the *Critique* seems to play an important part in Kant's account of perception. Nor, perhaps, is it sensitive enough to the subtleties of the ways in which the imagination of non-actual perceptions enters into our actual perception of something as an object, or as an object of a particular kind; subtleties which were recognized by Kant, and which are explored by Strawson in his paper 'Imagination and Perception', included in this volume.

H. E. Matthews ascribes to Strawson the view that 'transcendental idealism equals phenomenalism plus noumenalism' – noumenalism being the doctrine that there is a real though unknowable world of things in themselves as well as the world of appearances. Whether this is entirely fair to Strawson does not matter, but Matthews offers a different account of transcendental idealism, which seeks to make the contrast between the two worlds harmless and unobjectionable. In his essay on the subject, also in this volume, he maintains that when Kant talks about appearances he is not talking about a special kind of thing which our minds are somehow responsible for making, but is rather talking about things from a particular point of view – the human point of view. There are, or at least there could be, alternative points of view available to beings of other kinds; God's way of knowing things may be very different from ours, for instance. Indeed we ourselves adopt a different point of view when we think of ourselves as agents from the one we take up when we are concerned with knowing and understanding the world around us. Nothing can be said about the world as it is 'in itself', independently of any standpoint; the only function of such talk is to express the fact that these various standpoints are all standpoints on the *same* world.

This is an interesting suggestion, I think. But two questions arise. First, when Kant says that the world of appearances is constructed by us from the data of intuition, in accordance with the categories and the forms of space and time, is he really saying no more than Matthews does when he says that in seeing things as arranged spatially, temporally, causally and so on we are seeing things from one particular point of view? Part of the difficulty in answering this is that it is not wholly clear what Matthews' suggestion involves. It sounds harmless, and it is designed to be, but I am not sure that if it were fully spelt out those who lack sympathy with Kant would find it unobjectionable. Secondly, and perhaps more importantly, has Matthews captured all that Kant means by talking of things in themselves? It is not indeed obvious that his own position is a stable one, for it is not obvious that he can assign a definite sense to the claim that different standpoints represent the same world. He is aware of this objection, and considers it on p. 147, though some may doubt whether he has answered it fully.

It is certainly natural to feel that Matthews is trying to turn transcendental idealism into too harmless, too respectable a doctrine. But at the same time one may have doubts about dismissing Kant's talk of things in themselves and his two-worlds doctrine as just a lapse into the kind of metaphysics he was most concerned to reject. One is then left with a question which is not easy to resolve but which is well worth resolving: just how, precisely, *should* transcendental idealism be formulated?

* * *

Two of the papers in this collection are principally concerned with the transcendental deduction of the categories, Henrich's and Strawson's. Henrich's article discusses the structure and purpose of the argument as a whole; Strawson's does not, but helps to illuminate one of the fundamental and most difficult of its concepts, the concept of synthesis.

By no stretch of the imagination is the transcendental deduction easy to follow, and it has not been made easier by the propensity of many commentators to impose upon it interpretations which have little immediate connection with what Kant actually says. Henrich directs our attention back to the text, and particularly to the text of the second edition; the argument of the first edition, he thinks, is vitiated by an ambiguity in what is meant by calling an experience 'mine', an ambiguity of which Kant later became suspicious (though even in the second edition he is still not fully clear about it). Whether or not one wholly agrees with Henrich about this, it is evidently sensible to follow him in concentrating primarily on the second edition deduction, partly because those who have tried the alternative approach have not been strikingly successful, but mainly because, after all, Kant himself must clearly have thought his second version an improvement on his first. Now the second edition deduction is divided into two parts in a rather puzzling way; just when it looks as though the argument has been completed Kant tells us that it has not been and that there is more to come, the extra element being supplied in section 26. The first part of the proof shows that in so far as our intuitions are united they are subject to the categories. The second part extends the argument to all our intuitions, through the consideration that the pure intuitions of space and time must themselves exhibit a category-governed unity. Unlike most commentators, but again quite clearly rightly, Henrich takes this two-part structure seriously, and indeed relates it to the ambiguity in 'mine' in the first edition. He does not by any means solve all the problems of the transcendental deduction, and one may contest some of the details of his account; one may also disagree with his claim that the resulting proof is 'convincing in the context of Kant's philosophy' (p. 78). What is, I think, difficult to dispute is that he has helped to make clearer what the basic framework of the argument must be.

One thing Henrich does not discuss is the nature of synthesis. If one understood fully what Kant means when he speaks of the synthetic activity of the mind, one would understand in just what sense he thinks the phenomenal world is our construction. What he says about synthesis in general is not on the whole very helpful, however, so it makes good sense to start by looking at the things he says about particular kinds of synthesis and how they work. By seeing how it operates in specific types of case we may hope to gain a better picture of what synthesis really is.

He says a number of puzzling things about the transcendental synthesis of the imagination and the part it plays in perceptual recognition. It is to the elucidation of these that Strawson's paper is directed. Although initially they sound cryptic and obscure, Strawson considers that they make a substantial contribution to the philosophy of perception. To recognize something as a particular individual thing, or to see it as a thing of a familiar kind, involves relating the present perception to other actual or possible perceptions of the same thing or of other similar things. These other perceptions are not consciously called to mind, but nevertheless there seems to be a good sense in which they are alive in the present perception. This bringing together of perceptions, which does not require us to be aware of each of them separately, is certainly an important part of Kant's conception of synthesis. And Strawson argues that in calling it a synthesis of the *imagination* he is not giving this word a new and technical meaning, but only exploiting a natural analogy with the cases in which we more normally use it – an analogy which also struck Wittgenstein.

The question *how* the categories enter into our construction of the phenomenal world is considered by Kant in his short but highly problematic chapter on schematism. The chapter is problematic partly because it is not immediately obvious what question he is trying to answer. He complains about a lack of homogeneity between the categories and the empirical intuitions to which we must apply them, and says this makes it difficult to see how their application is possible. But many commentators have failed to see how there can be a difficulty here at all. The case was put most forcibly by G. J. Warnock,[7] who argued that Kant was simply confused as to what is involved in possessing a concept. To possess a concept is to know how to use it, and in particular how to apply it. There cannot therefore be any problem of how the categories are to be applied to experience once we know what the categories are. Homogeneity, whatever it amounts to, has nothing to do with the matter.

Chipman's paper is a defence of Kant against this kind of criticism, and a useful discussion of what the schematism chapter achieves. When Kant says that the categories are heterogeneous from empirical intuitions what he means is that it is not in virtue of any immediately recognizable phenomenal characteristic, like redness or triangularity, that they apply to them; nor in virtue of any set of such characteristics either. It is therefore legitimate to ask how we do subsume appearances under these concepts, and the question will be answered by finding the categories' schemata, for schemata are rules for the application of concepts. To say as in effect Warnock does that one cannot separate a concept from its schema is quite untrue, for even among empirical concepts there are many that one can possess without knowing how to recognize instances of them. The concept of bone marrow would be an example; or any concept with a substantial theoretical content. What Kant himself says in the chapter about how the schemata of the categories are to be found is not very satisfactory, and indeed Chipman thinks it is mostly worthless. But in a way that does not very much matter, for it is not here but in the remainder of the Analytic of Principles that the real work has to be done. There Kant examines the conditions for spatio-temporal experience, and shows in turn (or tries to) that each of the categories must be applied in a particular way. The function of the schematism chapter is really to act as a sort of prologue to that discussion. This last is an important point, and it deserves more emphasis than Chipman gives it. If the schemata were really meant to be derived in this chapter the rest of the Analytic of Principles would be very largely redundant; and it would be an anomalous addition, in view of Kant's insistence later on in the book that there cannot be two separate transcendental proofs of the same proposition.[8] For claims such as 'In all change of appearances substance is permanent' would then be proved in two quite distinct ways, once directly and once by the longer route of the deduction and schematism.

The paper by Walsh is an extract from his book *Reason and Experience*.[9] It examines Kant's account of self-knowledge, setting it in the context of the alternative views which Walsh ascribes to rationalists and empiricists. Kant agrees with the empiricists in his attack on the idea that knowledge of the active self is to be gained by rational intuition, but like the rationalists he is willing to recognize an active self which is the subject of experience and which is not an object of empirical knowledge. This active self can (and must) be '*thought* as subject', but it cannot be known. Walsh is more sympathetic to Kant than most writers are, and defends him from a number of criticisms, both from the empiricist and from the rationalist side. He does, however, argue that although he did not recognize it Kant's own position commits him to the existence of a power of non-empirical self-awareness, whereby reason is able to discover certain truths about its own nature, and upon which the possibility of serious philosophy – including that of the *Critique* – depends. He maintains furthermore that such a view of the matter is not only desirable on grounds internal to Kant's own theories, but also substantially correct in its own right. To this unfashionable but stimulating suggestion it would seem reasonable, in view of what has been said above, to add another observation. Truths discovered in this way would be both synthetic and *a priori*, and it would seem impossible that their truth should be read into the world by our minds. But it does not seem wholly impossible that they might be established as true by means of transcendental arguments, though finding such arguments might not be an easy task. Much of the obscurity and confusion of Kant's account of self-knowledge is due to his failure to see that there is no reason in principle why transcendental arguments should not establish conclusions which are about the real world (to which the active subject belongs), and are not confined to that phenomenal reality which we ourselves construct.

I have already mentioned Bennett's essay, as an application of his phenomenalistic interpretation of Kant. Bennett suggests that the reason why Kant thought there was something wrong with the claim that the world extends infinitely in space and in past time was that he did not think it susceptible of a phenomenalistic translation. Under such a translation, statements about the extent of the world become statements about the experiences we should have if we set about exploring it; if we had to translate the statement that the world extends infinitely, we could do so only by saying that the series of explorations would never be complete, however far it was carried. As a matter of fact this translation would be perfectly all right. But Kant, who was confused about infinity, did not think it would be. He thought there was a difference between saying that a series is infinite and saying that however far one goes along it one can still go further. So he thought that a claim to the effect that the world is actually infinite could not be handled phenomenalistically, and thus could not be a claim about the world of appearances.

It is clear, I think, that Kant did make this mistake about infinity, and that it affected his treatment of the antinomies. It was a pretty common confusion in those days. It is less clear, as I have said, that he is committed to the kind of phenomenalism Bennett ascribes to him. But even if he is not, Bennett may be essentially right about the present case. For he does hold that the world of appearances is a construction which is in some sense due to us; it is enough for present purposes if he is committed (as he presumably must be) to the view that empirical totalities are constructions out of the constituents of those totalities, for then the claims about the infinity of the world will have to be handled along very much the lines Bennett indicates.

So far I have not mentioned Kant's views about mathematics. But the first two papers are devoted to these. Parsons is mainly concerned with arithmetic, and with trying to discover what Kant's reasons were for calling arithmetical truths synthetic. He opposes a suggestion which has been argued forcibly by Hintikka,[10] to the effect that Kant regarded statements as synthetic when their proof requires a step in which something is first of all established by considering a particular case chosen arbitrarily, and is then said to hold more generally in view of the fact that the particular case was arbitrarily chosen – the kind of step that some logic books call existential instantiation or existential elimination. Hintikka thinks this is why Kant thought mathematics involved intuition, for when Kant spoke of intuitions he meant particular representations as opposed to general concepts. Parsons disputes the adequacy of this account of intuition, as a number of writers have: Kant also required that the object of intuition be directly present to the mind. In Parsons' own opinion, Kant appealed to the forms of intuition in order to verify the existential assumptions that mathematicians make. The concrete figures that the geometer constructs provide instances of what is under discussion and so establish its reality; in the same way, our ability to write down the series of numerals, and symbolically to express numerical operations, establishes the reality of the numbers and their relationships. This is an interesting idea, whether or not it is entirely correct – and the fact that, as Parsons himself points out, similar considerations would apply to logic too may provide some reason for hesitating about it. One may indeed wonder, after reading this careful study, whether any clear and comprehensible account of Kant's views on arithmetic can be given. What he says on the subject is suggestive and it has had much influence, but it does seem possible that he may have been rather confused.

Hopkins discusses geometry, where Kant's position seems a good deal more attractive. In *The Bounds of Sense*[11] Strawson attempted a limited defence of him, arguing that it is not merely an empirical matter that the space of our visual imagination is Euclidean. One of Hopkins' criticisms requires us to substitute 'approximately Euclidean' for 'Euclidean' here, since we can neither see nor imagine things in such microscopic detail as to be able to say they are not in a space which is only very slightly different from Euclidean space; but that is perhaps a minor point. It does seem to be clear that we cannot visualize two distinguishable straight lines joining the same two points, for example. And this is not because we have made it analytic by covertly building it into the meaning of 'straight' that no two such lines can both be straight, for as Hopkins says that would not tell us which of them was curved, and in any case we are not inclined to regard the suggestion that there might be two such lines as actually self-contradictory. Is it then synthetic *a priori*? Hopkins argues not. In his view it is indeed synthetic, but it is a mistake to regard it as non-empirical. The mistake is a natural one, for our imaginations are limited by what we have experienced, and none of us has witnessed two visually distinct lines, both observably straight, joining the same two points. This limitation on our imagination does not mean that others cannot have experienced such things and be able to imagine them freely; indeed, it is conceivable that we might ourselves come to be able to. A person blind from birth may be wholly unable to imagine colours, yet such people can sometimes be given sight and can then experience them and imagine them as readily as anyone.

Plausible though this may sound, I find it hard to be convinced that there is nothing more to it. I feel, in fact, as confident that I shall never witness two such lines, as I do that I shall never come across some state of affairs that is actually self-contradictory; though in this case the possibility does not seem to be ruled out by logic. There is a parallel, I think, with 'Nothing is both red and green all over at the same time' – another claim one can be sure of *a priori,* though again it is arguably not an analytic truth. The matter is a complicated one, but it does look as though we might have here examples of real synthetic *a priori* knowledge. It seems very doubtful, however, that it can be accounted for satisfactorily in either of Kant's two ways. One is reluctant to say that it is imposed by our minds upon the world of appearances, and we cannot reasonably hope to vindicate it by transcendental arguments, for it would be difficult indeed to argue that experience would be impossible if things were noticeably non-Euclidean, or red and green all over. So, at least in this area, it looks as though we may be left with the still unsolved problem: how *is* synthetic *a priori* knowledge possible?[12]

[1] As Walsh for example believes: see p. 164 below.
[2] London: Methuen, 1959.
[3] London: Methuen, 1966.
[4] Cf. E. Adickes, *Kant und das Ding an Sich* (Berlin: Pan, 1924), esp. ch. 1.
[5] *Prolegomena,* sect. 36; *Critique of Pure Reason,* A 127.
[6] *Kant's Analytic* (Cambridge University Press, 1966), 23 ff., 127.
[7] 'Concepts and Schematism', *Analysis* IX (1948–9).
[8] A 787 f./B 815 f.
[9] Oxford: Clarendon Press, 1947.
[10] For references see the Bibliography and Parsons's footnotes.
[11] Part five.
[12] I am very grateful to John Kenyon for reading a draft of this Introduction, and for making a number of valuable suggestions.

I
KANT'S PHILOSOPHY OF ARITHMETIC[1]
CHARLES PARSONS

THE interest and influence of Kant's philosophy as a whole have certainly been great enough so that this by itself would be enough to make Kant's philosophy of arithmetic of interest to historical scholars. It is also possible to show the influence of Kant on a number of important later writers on the foundations of mathematics, so that Kant has importance specifically as a figure in the history of the philosophy of mathematics. However, my own interest in this subject has been animated by the conviction that even today what Kant has to say about mathematics, and arithmetic in particular, is of interest to the philosopher and not merely to the historian of philosophy. However, I do not know how much of an argument the following will be for this.

Kant does not discuss the philosophy of arithmetic at any great length, so that it is virtually impossible to understand him without making use of other material. What I have used consists mainly of two considerations: the integration of Kant's theoretical philosophy as a whole, and modern knowledge on the foundations of logic and mathematics. The justification for using the second is twofold; first, I think experience shows that one does not get far in understanding a philosopher unless one tries to think through the problems on their own merits, and in this one must use what one knows; second, if one is today to take Kant seriously as a philosopher of mathematics, one must confront him with this modern knowledge, which after all in major respects shows immense progress from the situation in his lifetime.

I shall be concentrating mainly on one question, which I think must be answered before one goes farther with the subject: why did Kant hold that arithmetic depends on sensible intuition, indeed that arithmetical propositions in some way refer to sensible intuition? This is, of course, closely related to the question of why he regarded such propositions as synthetic rather than analytic. In considering this question, one must very soon consider Kant's views on logic and its relation to arithmetic. Also since the answer to the above question is much clearer if 'arithmetic' is replaced by 'geometry,' we shall also give some consideration to Kant's views on geometry.

From Morgenbesser, Suppes, and White (eds.), *Philosophy Science, and Method. Essays in Honour of Ernest Nagel*. London and New York: Macmillan and St Martin's Press, 1971, pp. 568–94. © 1969, 1971 by St. Martin's Press and used with permission.

In order to clarify our problem, let us first briefly consider Kant's concept of intuition. Intuition is a species of representation (*Vorstellung*) or, in the language of Descartes and Locke, idea. Having intuitions is one of the primary ways in which the mind can relate to or be conscious of objects. The nearest thing to a definition in the *Critique of Pure Reason* occurs in a classification of representations:

This [knowledge] is either *intuition* or *concept* (*intuitus vel conceptus*). The former relates immediately to the object and is singular [*einzeln*], the latter refers to it mediately by means of a feature which several things may have in common. [A 320/ B 376–7.][2]

In the opening sentence of the Transcendental Aesthetic, Kant says:

In whatever manner and by whatever means a mode of knowledge may relate to objects, *intuition* is that through which it is in immediate relation to them, and to which all thought as a means is directed. [A 19/B 33.]

A passage in §1 of Jäsche's edition of Kant's lectures on Logic reads:

All modes of knowledge, that is, all representations related to an object with consciousness are either *intuitions* or *concepts.* The intuition is a singular representation (*repraesentatio singularis*), the concept a general (*repraesentatio per notas communes*) or *reflected* representation (*repraesentatio discursiva*).[3]

Intuitions are thus contrasted with *concepts,* which relate to objects only mediately, by way of certain properties and by way of intuitions which instantiate them and which relate indifferently to all the objects which possess the required properties.

What is meant by calling an intuition a singular representation seems quite clear. It can have only one individuated object. The objects to which a concept 'relates' are evidently those which fall under it, and these can be any which have the property which the concept represents, so that a concept will only in exceptional cases have a single object. Thus far, the distinction corresponds to that between singular and general terms.

One might think that the criterion of 'immediate relation to objects' for being an intuition is just an obscure formulation of the singularity condition. But it evidently means that the object of an intuition is in some way directly present to the mind, as in perception, and that intuition is thus a source, ultimately the only source, of immediate knowledge of objects. Thus the fact that mathematics is based on intuition implies that it is immediate knowledge and thus, even though synthetic *a priori*, does not require the elaborate justificatory argument which the Principles do (A 87/B 120). By the immediacy criterion Kant's conception of intuition resembles Descartes's, while by the singularity criterion and his insistence on a non-intuitive conceptual factor in all knowledge, Kant's theory of intuition differs from that of Descartes.

That what is immediately present to the mind are individual objects seems to be an axiom of Kant's epistemology, or one might also say metaphysics, since it goes with the conviction that objects, the primary existences, are in the first instance individual objects. Thus what satisfies the immediacy criterion of intuition will also satisfy the singularity criterion.

It does not seem that the converse must be true. The idea of a singular representation formed from concepts seems quite natural to us. Such a representation would relate to a single object if to any at all, but it hardly seems immediately. By associating it with a definite description rather than with a general term, we would distinguish it from a concept under which exactly one object falls (even if necessarily). For Kant, however, the passage from A 320/B 376–7 seems to allow such a representation to be a concept; this might also be suggested by the fact that the idea of God is called a concept; it is nowhere suggested that it is an intuition. However, Kant never remarks, so far as I know, on the implications of the possibility of non-immediate singular representations for the concept of intuition.

This omission may give support to a theory which has been advanced by Jaakko Hintikka according to which the singularity criterion is the sole defining criterion: An intuition is simply an individual representation.

In Kant and in his immediate predecessors, the term 'intuition' did not necessarily have anything to do with appeal to imagination or to direct perceptual evidence. In the form of a paradox, we may perhaps say that the 'intuitions' Kant contemplated were not necessarily very 'intuitive'. For Kant, an intuition is simply anything which represents or stands for an individual object as distinguished from general concepts.[4]

Many of the passages Hintikka cites also mention the immediacy criterion, and it is not clear why Hintikka thinks it non-essential. The main reason, which we shall consider later, is that this assumption supports a theory of Beth and Hintikka to explain Kant's notion of 'construction of concepts in intuition' and the resulting analysis of mathematical demonstration Another seems to be the absence of the immediacy criterion in the *Logic* and the fact that Kant makes remarks on concepts which seem to exclude essentially singular concepts and thus to imply that all singular representations are intuitions.[5]

Hintikka also points out that the part of the Trancendental Aesthetic where Kant argues that space is an intuition argues essentially that the representation of space is singular. However, he has opened the Aesthetic by stating the immediacy criterion (A 19/B 33, cited above) and in the proof of intuitivity he does say that space is *given* (B 39, also A 25). Moreover, in arguments for the same thesis in the Inaugural Dissertation of 1770, Kant does appeal to immediacy: immediately after arguing that space is a pure intuition because it is a 'singular concept', Kant says of geometrical propositions that they 'cannot be derived from any universal notion of space but only as it were *seen* in space itself as if in something concrete' (§15c, emphasis Kant's). Later he says, 'Geometry makes use of principles which fall under the gaze of the mind.'

It seems to me that the textual evidence for Hintikka's view is not sufficient to outweigh the clear statements and emphases on the immediacy criterion, even though the alternative view must assume that Kant in discussing these matters did not keep in mind the possibility of non-immediate singular representations.[6] But Hintikka's theory really stands or falls on the interpretation of the role of intuition in mathematics.

A thesis about intuition which is of great importance for Kant is that our mind can acquire intuitions of actual objects only by being *affected* by them. Just what this 'affection' is I shall not venture to say, but it involves for the subject a certain passivity, so that our perceptions are not on the face of it brought about by our own mental activity, and also a certain exposure to contingency in our relations with objects. Thus we do not perceive objects unless they physically affect our sense organs.

A particular and highly important twist of Kant's philosophy is that the nature of our capacity to be affected by objects, our *sensibility,* already determines certain characteristics of our intuitions. These are said to be the form of our intuition in general. Among them is *spatio-temporality.* This must be understood to mean that the nature of the mind determines that the objects we intuit should be spatial and temporal, and indeed intuited as such. The intuition which plays a role in mathematics, which is not the direct result of the affection of our mind by objects, expresses an intuitive insight which we have into our forms of intuition and is in that sense still an intuition of sensibility. It is apparently also sensible intuition in the sense of being intuition of *inner* sense.

As Hintikka rightly emphasizes, this intrinsic connection between intuition and sensibility does not come directly from the concept of intuition but represents a characteristic of man, or more generally of finite intelligences. Such an intelligence derives the content of its consciousness from outside with the resulting exposure to contingency and the necessity of concepts in order to represent objects not present. Thus not only sensibility but also thinking, or consciousness through concepts (knowledge through concepts, A 69/B 94), are characteristics of finite intelligences. The alternative is an 'intuitive understanding' whose activity would *create* the objects of its awareness. Its awareness would be *only* intuition; it is called 'intellectual intuition' because it has the spontaneity which for us is characteristic of thought and because the unity which with us is the result of synthesis of the *given* is for it already present in the intuition. It seems clear that intellectual intuition would satisfy the immediacy criterion.

Let us now turn to Kant's views on logic. What must strike a person with modern training most forcefully in considering Kant's outlook on logic is the limitation of his knowledge of and conception of it. Kant learned and taught the established logical lore at a very uncreative time in the history of the subject. Thus the formal logical analysis he undertakes is pretty well limited to the categorical proposition-forms of the theory of the syllogism, with gestures toward hypothetical and disjunctive propositions. The inferences which are covered are the syllogisms and immediate inferences of the Aristotelian theory and a few propositional inferences such as *modus ponens*. Of propositional logic as an additional developed theory, or of the additional possibilities of quantification theory, Kant had no idea.

Kant not only had very limited technical resources at his command; what is more striking and more damaging to his standing as a philosopher, he was largely satisfied with logic as he found it. Technically he could hardly in any case have gone very far beyond the state of the science in his own time, and he was not a creative mathematician. But what would have been needed for Kant to be dissatisfied with 'traditional logic' might only have been more insight into his own discoveries.

As is well known, Kant attributed the lack of progress in logic to the absence of any need for it. He held that logic was established as a science and then finished off once and for all by Aristotle. This is a false view not only of the possibilities of discovery in logic but also of the history of the subject, which, far from not being 'able to advance a single step' nor 'required to retrace a single step' since Aristotle, had done both more than once. Kant's opinion was also influential and served to create resistance to more reasonable views both of logic itself and of its history.

Why Kant should have thought the science of logic both completable and completed is a question which I shall not attempt to answer here. I do not know whether a serious effort to answer it would uncover interesting ideas of Kant, which as it is we do not understand. In general, it can be said that the view harmonized extremely well with the more rationalistic side of Kant's way of thinking and with the belief, which he was not the only great philosopher to hold, that his own work finished off an important part of philosophy. Kant certainly thought that there were inexhaustible sources of problems, even philosophical problems, for the human intellect to wrestle with. But he held that this inexhaustibility lay within limits fixed by a form, the basic properties of which could be exhaustively described. This form would belong to the human faculty of thought itself, which so long as it was dealing with 'itself and its form' and not with objects given from outside or with the manner in which they might be given from outside, was bound to be capable not only of being on sure ground but of uncovering and analyzing every relevant factor. Reason and the understanding are 'perfect unities'. (A xiii, A67/B92.) We also find an echo of the Cartesian idea that the self is better known than objects:

I have to deal with nothing save reason itself and its pure thinking; and to obtain complete knowledge of these, there is no need to go far afield, since I come upon them in my own self. [A xiv.]

Logic is, according to Kant, the most general of all divisions of knowledge. It applies to all objects of our thought in general, and all true statements and sound inferences must conform to it. In particular and especially important, logical possibility is the most inclusive kind of possibility. If something is possible in any respect whatsoever, it is logically possible; its concept does not involve a contradiction. In particular, at least as far as Kant's explicit statements are concerned, the applicability of logic is not limited by the forms of our sensibility.

The relation between logic and the forms of intuition can best be seen by contrast with geometry. The forms of intuition provide the basis for certain necessary truths, in particular those of geometry, in the sense that if the forms of intuition were not as they are the truths in question would not hold, and if we did not have a certain insight into our forms of intuition, we would not know them. The application of these truths, however, is limited to the objects which affect our senses. Moreover, the principles are true of these objects only as they appear and not as they are in themselves.

These limitations do not obtain for logic. In particular, there are states of affairs which are logically possible but which are excluded by the forms of intuition, such as the existence of spatial configurations contrary to the theorems of Euclidean geometry; so that geometry is a more special theory than logic, not only in the sense that it deals with a more restricted type of object but also in the sense that it makes statements about these objects which are not logically necessary, although they are necessary in another way.

Logic is also not subject to the great limitation of knowledge based on intuition, that of appearances. When Kant says that it must be possible to *think* of things in themselves, he implies first that such a conception does not contradict the laws of logic, and second that in the statements we make about them, the logical laws are still a negative criterion of truth. If he could not trust logic in this realm, Kant's metaphysics of morals would not be able to get off the ground.

Already on this level, it is possible to see quite clearly some reasons why

Kant should have regarded geometry as synthetic *a priori* and used an idea such as that of a form of intuition in order to explain how such a science was possible. Geometry is a more special theory than logic first in the sense that it contains non-logical primitives, second in that its theorems cannot in general be proved merely by means of definitions and logic, as Leibniz apparently thought. Indeed this is much more obvious to us than in Kant's time, given that we have non-Euclidean geometry and are in general less tempted to overestimate the power of logic, especially traditional logic. It is worth pointing out that Euclid's postulates are what are in effect existence assumptions, so that here Kant's general views about existence would imply that they could not be analytic.

That Kant should then found geometry on the form of our sensible intuition is not difficult to understand. On the one hand spatio-temporality is a characteristic property of the objects given to the senses. Moreover, Kant emphasized that space was an individual the notion of which was understood in a way analogous to ostension, and the same ostensive under-standing would be necessary for the particular primitives of geometry. On the other hand, Kant started from the idea that geometry was a body of necessary truths with evident foundations. That the axioms of geometry should be empirically verified directly is contrary to their necessity; that they should be some sort of high-level hypotheses is contrary to their evidence.

The second observation to make about Kant's views on logic is that he never suggests a conventionalist account of logical validity. It is true that the very general character of logical and analytical truths goes with their uninformativeness. They reflect the nature of the mind and of certain particular concepts, and apparently not at all how the world is otherwise. But this nature and the manner in which particular concepts give rise to the analytic truths that they do seem to be something given, which will be in fact the same for all discursive intelligences, even if their forms of intuition are quite different from ours.

Kant does not give much explanation of how this is, and perhaps he felt some doubt as to the possibility of giving such an explanation. If we try to apply the insight which we might get from the 'Transcendental Deduction' to this question, we get into a very difficult dilemma. Namely the essential activity of the understanding seems to be in relation to material given in intuition, to bring it to the unity expressed in an objective judgement. In other words, the notions of object, concept, judgement get their whole sense from their application to experience. Nonetheless the understanding has a greater generality than intuition. The forms of intuition are not logically necessary; and in operating logically with a given notion, it is not necessary to appeal to intuition or even to suppose that the notion has an intuition corresponding to it. It is possible in some way for us to recognize that what can be given in experience is not the whole of possible reality, and even to recognize that with the help of intuition we ca⌐ know objects only in a relative way, as they appear. All this points, even apart from the requirements of Kant's moral philosophy, to the presence in us of more general conceptions of object, concept, judgement, and *a fortiori* inference. This dilemma will occupy us again later, since it has an application to the problems of arithmetic.

With respect to Kant's philosophy of geometry, the difficulties do not concern why Kant thought geometry to be *a priori* intuitive knowledge, but rather whether this is true and what precisely the theory was by which he proposed to explain how it could be true. When we turn to Kant's philosophy of arithmetic, there is even less difficulty as to why he should have thought arithmetical propositions *a priori*. But it is already by no means easy to see why Kant regarded them as synthetic, as based in some way on our forms of intuition, in particular on the form of inner intuition, time, and as limited in their application to appearances.

It will become clearer why Kant regarded arithmetical propositions as *synthetic* if we observe that Kant's concept of analytic proposition most likely had a much narrower extension than the corresponding concept in more recent philosophy, e.g., in Frege and logical positivism. Kant does not formulate his concept with enough precision so that we can be altogether sure about this. But it seems rather clear from the examples that when Kant speaks of the concept of the predicate of an analytic judgement as *contained* in that of the subject, the situation is analogous to that in which the subject concept is defined by the conjunction of the predicate concept with perhaps certain others. This would be a paradigm case where the connection of subject and predicate is 'thought through identity' (A 7/B 10). An idealized version of an analytic judgement would be one of the form 'All *AB* are *A*', or 'All *C* are *A*', where '*C*' is defined as '*A* and *B*'. This is idealized because, according to Kant, outside mathematics concepts do not in general have definitions in the proper sense.

It seems certain that a number of other forms would have to be admitted as analytic, e.g., 'No *AB* are not *A*' or the propositional 'If *p* and *q*, then *p*'.[7] But there is no particular reason why '7 + 5 = 12' should be. Kant says (B 15) that for '7 + 5 = 12' to be analytic, it would have to follow from the concept of a sum of 7 and 5 by the law of contradiction. This would be as if it were provable from definitions by a very restricted logic, probably included in the limited traditional apparatus at Kant's command, and it is hard to see how it could be true otherwise.

However, it is one thing to say there is no reason to expect this and another to understand Kant's specific reason for thinking it false. Kant indicates that the way you find out that $7 + 5 = 12$ is by a process like counting, of progressing from 7 to 12 by successive additions of 1, in which one must operate with a *particular instance* of a group of 5 objects, which can only be given in intuition.

We have to go outside these concepts, and call in the aid of the intuition which corresponds to one of them, our five fingers, for instance, or, as Segner does in his *Arithmetic*, five points, adding to the concept of 7, unit by unit, the five given in intuition. For starting with the number 7, and for the concept of 5 calling in the aid of the fingers of my hand as intuition, I now add one by one to the number 7 the units which I previously took together to form the number 5, and with the aid of that figure [the hand] see the number 12 come into being. [B 15–16.]

It is, however, still not clear why that process cannot be either itself put in the form of a purely logical argument or replaced by something quite different which can.

There was an attempt to do just this with which Kant was in a position to be familiar, by Leibniz in the *Nouveaux Essais*.[8] Leibniz worked with '$2 + 2 = 4$', but the type of argument suffices for any addition formula. He assumed as an axiom the substitutivity of identity, which Kant would in all probability have regarded as analytic. Leibniz took as definitions

$$2 = 1 + 1,$$
$$3 = 2 + 1,$$
$$4 = 3 + 1,$$

which is approximately what is done in modern formalizations. Then the proof goes as follows:

$$
\begin{aligned}
2 + 2 &= 2 + 1 + 1 &&\text{(def. of '2')}\\
&= 3 + 1 &&\text{(def. of '3')}\\
&= 4 &&\text{(def. of '4')}
\end{aligned}
$$

The standard modern objection to this argument is that Leibniz should have inserted brackets, so that it goes

$$2 + 2 = 2 + (1 + 1) = (2 + 1) + 1 = 3 + 1$$

and therefore assumes an instance of associativity. We cannot exclude the possibility that this was known to Kant when he was working on the *Critique of Pure Reason*, since it occurs in effect in the book *Prüfung der kantischen Kritik der reinen Vernunft*, vol. I (Königsberg, 1789), by Kant's pupil Johann Schultz, professor of mathematics in Königsberg.

Putting great weight on the evidence of writings by Schultz and other disciples of Kant, Gottfried Martin has put forth the hypothesis that Kant envisaged an axiomatic foundation of arithmetic similar to the classical axiomatizations of geometry.[9] He sees the claim that arithmetic is synthetic as resting on the first instance on the logical point that arithmetical propositions such as '7 + 5 = 12' cannot be proved by mere logic from definitions such as those Leibniz uses. An axiomatic foundation of the sort which would answer to Martin's ideas is given in Schultz's *Prüfung*. Without explicitly mentioning Leibniz, Schultz points out that the sort of proof of an arithmetical identity that Leibniz gives rests on the assumption of associativity. He gives, for '7 + 5 = 12', an argument which also rests on commutativity, and seems, wrongly, to think this assumption unavoidable. But of course commutativity has to be used sooner or later in arithmetic.[10]

Schultz gives two axioms, the commutativity and associativity of addition. He neither asserts nor denies the independence of the corresponding laws of multiplication and of the distributive law. He also gives two 'postulates' which are worth quoting in full:

1. From several given homogenous quanta, to generate the concept of one quantum *by their successive connection*, i.e., to transform them into *one whole*.
2. To *increase* and to *diminish* any given quantum *by as much as one wants*, that is, *to infinity*.[11]

The second postulate implies that Schultz is not thinking specifically of the arithmetic of integers but also of continuous quantities. In any case, the first postulate is the basis for the supposition that the function of addition is *defined*; i.e., given numbers m, n, there actually exists a number $m + n$.

If we accepted this as actually giving Kant's conception, there would still remain the question how intuition enters into the foundation of these axioms and postulates. About this Schultz has in fact something to say. But in transferring the conception to Kant we are faced immediately with the difficulty that he explicitly says that arithmetic does not have axioms.

As regards magnitude (*quantitas*), that is, as regards the answer to be given to the question, 'What is the magnitude of a thing?' there are no axioms in the strict meaning of the term, although there are a number of propositions which are synthetic and immediately certain (*indemonstrabilia*). [A 163–4/B 204.]

He considers two possibilities, rules of equality, which he asserts to be analytic (a proper axiom must be synthetic), and the elementary arithmetical identities, such as '7 + 5 = 12', which are what he seems to be referring to at the end of our quotation, which are indeed synthetic and indemonstrable, but which he declines to call axioms because they are singular.

This position is reaffirmed in a letter from Kant to Schultz dated November 25 1788,[12] in which he comments on the manuscript of volume I of the *Prüfung*. There he gives a reason, which I shall mention later, why arithmetic should not have axioms. He does say that arithmetic has *postulates*, 'immediately certain practical judgements'. The general tone of his discussion suggests that he might regard the general directive to carry out addition, or the proposition that this can always be done, as postulates, i.e., that he might accept Schultz's first postulate. But what he seems to have specifically in mind is what he elsewhere calls numerical formulae, i.e., $7 + 5 = 12$.

We cannot be certain, however, that the mathematical material of the published version of the *Prüfung* was present in the manuscript that Kant was commenting on. For it seems from the letter, as Martin points out,[13] that the manuscript maintained that arithmetical propositions were analytic, and thus it is clear that it was considerably revised *after* Schultz received Kant's letter. The fact that in the published version the axiomatic analysis is used to support the conclusion that arithmetic is synthetic does not prove that it was not present in the manuscript, although the supposition that the *postulates* were there is a bit strained. But that Schultz might have argued that the commutative and associative laws were analytic is not at all impossible. (Leibniz argued this at least for commutativity.)[14]

Even so, unless one accepts Martin's rather unlikely idea that the axiomatic analysis was contributed by Kant to Schultz *after* the letter, it is hard to escape the conclusion that Schultz understood the mathematical issue in at least one respect better than Kant himself: Kant does not seem to have had an alternative view of the status of such propositions as the commutative and associative laws of addition. He can hardly have denied their truth, and it seems that if they are indemonstrable, they must be axioms; if they are demonstrable, they must have a proof of which he gives no indication.

If when speaking of the axiomatic character of arithmetic, Martin means that according to Kant arithmetic must make use of propositions which cannot be deduced by logic and definitions, then there can be no disagreeing with him. But if he means that Kant had in mind setting up arithmetic as an axiomatic system of which Schultz's is a very primitive instance and that it is in the verification of such laws as the commutative and associative that the primary application of intuition in arithmetic is to be found, then Kant's actual words go against him.

Even if Martin's view of the matter is quite correct as far as it goes, it cannot satisfy us. In the first place, it does not answer the question why arithmetic should depend on intuition, except in the sense, entirely bound to the primitive level of axiomatics in Kant's time, that so far as one can see the obvious alternative is insufficient. In the second place, it carries over to arithmetic the considerations which were at work in geometry while our original sense of difficulty arose from the difference between the two. And there are many indications, in particular some remarks in the letter to Schultz, which I shall discuss, that he saw some of this difference and did not intend to give an entirely symmetrical account.

The problem of the asymmetry of arithmetic and geometry could be solved by an interpretation suggested by E. W. Beth[15] and developed by Hintikka.[16] From their interpretation it seems to follow that if a proposition B of geometry is proved by a proof which appeals to axioms $A_1 \ldots A_n$ (I here include postulates),[17] then in general the conditional $A_1 \& \ldots \& A_n \supset B$ is synthetic; at any rate an appeal to intuition is made over and above any which is made in verifying the axioms. One could then argue that since arithmetic according to Kant does not have axioms, only the first type of appeal to pure intuition occurs in arithmetic.

Beth's and Hintikka's hypothesis is that for Kant certain arguments which can nowadays be formulated in first-order predicate logic involve an appeal to intuition. In view of the singularity criterion for intuition, the natural candidates for such arguments are arguments involving singular terms. For Beth the form of argument involved is illustrated by the proof that the base angles of an isosceles triangle are equal:

We proceed, as is well known, as a rule as follows: first we consider a particular triangle, say ABC, and suppose that AB = AC; then we show that $\angle$ ABC = $\angle$ ACB and have thus proved that the assertion holds in the particular case in question. Then one observes that the proof is correct for an arbitrary triangle, and therefore that the assertion must hold in general.[18]

The general form of the argument is as follows: We want to prove '(x) $(Fx \supset Gx)$'. We assume a particular a such that Fa. We then deduce 'Ga'. We then have '$Fa \supset Ga$' independently of the hypothesis. But since a was arbitrary, '(x) $(Fx \supset Gx)$' follows.

This form of argument, as for example in Beth's case, is the characteristic form of a proof in Euclid. In the Discipline of Pure Reason in Its Dogmatic Employment, the section of the *Critique* where Kant sets forth his view of mathematical proof as proceeding by 'construction of concepts in pure intuition', this form of argument appears clearly in the geometrical example (A 716–7/B 744–5). The geometer 'at once begins by constructing a triangle'. By a series of constructions on *this triangle* and applications of general theorems to it 'through a chain of inferences guided throughout by intuition he arrives at a fully evident and universally valid solution of the problem'.

Hintikka concentrates attention rather on the rule of existential instantiation, that is on arguments of the form where a is introduced to indicate an F, in view of the fact that the previous line affirms that there are Fs.[19]

Both of these arguments have in common that they turn on the use of a free variable which indicates *any* one of a given class of objects, so that an argument concerning it is valid for *all* objects of the class. They thus have a formal analogy with the appeal to pure intuition, in that a *singular* term is used in such a way that what is proved of it can be presumed generally valid. Moreover, the manner in which this generality is assured, namely by not allowing anything to be assumed about a except what is explicitly stated in premisses, is reminiscent of a statement of Kant about the role of a constructed figure in a proof:

If he is to know anything with *a priori* certainly he must not ascribe to the figure anything save what necessarily follows from what he has himself set into it in accordance with his concept. [Bxii.]

It is noteworthy that in traditional algebra calculations are carried out on terms and formulae with free variables, where the derivation of such an equation serves to prove a general proposition. Hintikka interprets the rather obscure remarks about 'symbolic construction' in algebra in this sense.[20]

It would naturally follow from the conception of intuition as simply individual representation that the mere form of these arguments is such that they involve intuition. Of course, it would not give any plausibility to Kant's more far-reaching philosophical theses which turn on the connection of mathematics with the form of *sensibility*. Thus the philosophically interesting aspects of the concept of pure intuition seem to lose their point when it is pointed out that these arguments can be formalized in pure quantification theory. This is exactly the conclusion which Beth draws.

One might object that this seems to presuppose that logic itself does not pose philosophical problems which the notion of pure intuition might be needed to answer, but on this at least Beth is in agreement with Kant in most of his utterances. But anyway it seems unlikely that the break between arguments which turn on the generality interpretation of free variables and logical arguments which do not is the philosophically most significant break within mathematical proof.[21]

One could wish more clear-cut evidence for the attribution of such a view to Kant or even for the most modest thesis that he started with this idea in developing his philosophy of mathematics. If it was his mature view, Kant's mathematically astute pupil Schultz seems not to have suspected it since there is no suggestion of it in the *Prüfung*. Schultz took for granted that an adequate axiomatization would be such that if the axioms were analytic so would be all the theorems. Mathematics fails to be analytic just because in its deductive development synthetic *premisses* must be used. The same view is expressed by Kant when he says:

For as it was found that all mathematical inferences proceed in accordance with [*nach*] the principle of contradiction (which the nature of all apodictic certainty requires), it was supposed that the fundamental propositions of the science can themselves be known to be true through that principle. This is an erroneous view. For though a synthetic proposition can indeed be discerned in accordance with the principle of contradiction, this can only be if another synthetic proposition is pre-supposed, and if it can then be apprehended as following from this other proposition. [B 14.]

Against this, it is pointed out[22] that Kant says of a geometric proof that it proceeds 'through a chain of inferences guided throughout by intuition' (A 716–7/B 744–5). In view of the description Kant gives of the proof, this could easily mean that in the course of the proof one is constantly appealing to the evidences formulated in the axioms and postulates. It would obviously be anachronistic to attribute to Kant a picture of proof modeled on a formal deduction where the axioms are stated at the beginning and everything else is logic and where the purpose is not to show the *truth* of the proposition proved but merely that it follows from the axioms. On the contrary, for Kant a Euclidean proof is convincing because on each particular application of an axiom or postulate the correctness of what it claims in this particular case is evident.

It must be conceded that it might be true that inference by certain rules from analytic premisses might yield analytic conclusions while inference according to the same rules from synthetic premisses could lead to conclusions which are not only themselves synthetic but such that the conditional of premisses and conclusion is also synthetic. In particular, the rule of existential instantiation can only come into play in the presence of an existential quantifier, and it is not clear that, for Kant, a statement in which an existential quantifier occurs essentially can be analytic. I can only say that in such cases the text of Kant does not clearly indicate that the necessity of an appeal to intuition arises for the *inference* and not merely for the verification of the premiss.

If Hintikka were right, one could expect that in the passages on algebra the role of variables would be emphasized. It is possible to find this emphasis in the passage on A 717/B 745, but it is not really explicit. The emphasis of A 734/B 762 seems different, where Kant says, 'The concepts attached to the symbols, especially concerning the *relations* of magnitudes, are presented in intuition.' (Emphasis mine.) The relations would seem to be expressed by algebraic function signs. Although the passages on algebra offer some support for Hintikka's theory, it is less than decisive. I shall show that there are other possible ways of looking at these passages.

The direct evidence thus seems to me on the whole opposed to the Beth–Hintikka theory. However, it would have strong indirect support if there were not other ways to explain how *arithmetic* can require pure intuition and to interpret the notion of 'construction of concepts', especially in algebra. To this end we now return to the problem of the difference between arithmetic and geometry.

The difficulty can be put in this way. The synthetic and intuitive character of geometry gets a considerable plausibility from the fact that geometry can naturally be viewed as a theory about actual space and figures constructed in it. This space is related to the senses by being a field in which the objects given to the senses appear, and geometry seems to give quite substantial information about this space which from the point of view of abstract thought might be false.

The content of arithmetic does not immediately suggest such a special character or such a connection with sensibility. Of course in the first instance it speaks of numbers and purely abstract operations and relations – equality, addition, subtraction, etc. Then the question is: what is the field of application of numbers? That is, what sorts of things can be counted, assigned cardinal or ordinal numbers, or measured and thus assigned continuous quantities? On the face of it, there is no reason to believe that the application of arithmetic need be to objects in space and time. Although this has certainly become more evident since the rise of abstract mathematics, that mathematical objects themselves could be numbered was something which Kant was certainly in a position to be aware of. If the application of arithmetic is to be limited to appearances, this limitation has to be understood rather broadly in order to reconcile it with obvious facts.

In the case of geometry, it was possible to mention logical possibilities which the concepts allowed but which did not exist according to the mathematical theory; Kant gives the example of a two-sided plane figure, and many more such possibilities were opened up in the development of non-Euclidean geometry. It was probably impossible in Kant's time to be clear about whether such a possibility exists in arithmetic. If it did, it would give rise to a clear separation of arithmetical from logical truth. This sort of argument was not available to Kant. The difficulty is made more acute, some would say insoluble, by subsequent developments in logic, particularly the efforts of Frege and others to do just what Kant thought impossible — to reduce arithmetic to logic, to deduce arithmetical propositions from definitions and propositions of pure logic.

Of course the extent of what counts as 'logic' here is considerably wider than what Kant regarded as such. At the very least, we need for this type of construction to incorporate some of the theory of classes into logic; not just the notion of class and some elementary operations concerning them, but also at least some modest axioms of class existence – how modest depending on how much arithmetic one wants to deduce.

Both to set forth what we need of this construction for our purposes and to indicate how far one can go without using set-theoretic devices, I shall discuss a logical truth which is closely related in meaning to '2 + 2 = 4' and provides the key to the proof of '2 + 2 = 4' in more extended formalisms. This example will help to indicate how far the cases of arithmetic and geometry are symmetrical.

Consider the following schema of the first-order predicate calculus with identity:

$$(\exists_2 x)Fx \cdot (\exists_2 x)Gx \cdot (x) - (Fx \cdot Gx) \cdot \supset (\exists_4 x)(Fx \vee Gx) \qquad (1)$$

where '$(\exists_0 x) Fx$' is an abbreviation for '$- (\exists x)Fx$' and '$(\exists_{n+1} x)Fx$' for

$$(\exists x)[Fx \cdot (\exists_n y)(Fy \cdot y \neq x)].$$

so that '$(\exists_2 x)Fx$' can be expanded as

$$(\exists x)(\exists y)[Fx \cdot Fy \cdot x \neq y \cdot (u)(Fu \supset \cdot u = x \vee u = y)] \qquad (2)$$

and '$(\exists_4 x)(Fx \vee Gx)$' as

$$(\exists x)(\exists y)(\exists z)(\exists w)[Fx \vee Gx \cdot Fy \vee Gy \cdot Fz \vee Gz \cdot Fw \vee Gw.$$
$$x \neq y \cdot x \neq z \cdot x \neq w \cdot y \neq z \cdot y \neq w \cdot z \neq w.$$
$$(u)(Fu \vee Gu \cdot \supset \cdot u = x \vee u = y \vee u = z \vee u = w)]. \qquad (3)$$

Intuitively, the proof of this schema goes like this: Suppose $(\exists_2 x)Fx$ and $(\exists_2 x)Gx$. Then in view of (2) and its counterpart for '$(\exists_2 x)Gx$' there are x, y, z, and w such that

$$Fx \cdot Fy \cdot x \neq y \cdot (u) (Fu \supset \cdot u = x \vee u = y)$$
$$Gz \cdot Gw \cdot z \neq w \cdot (u) (Gu \supset \cdot u = z \vee u = w).$$

We then go out to argue, with the help of '$(x) - (Fx \cdot Gx)$', that x, y, z, w satisfy the condition in the scope of (3), and so we infer that there *are* x, y, z, w such that this condition holds.

This schema requires for its formulation only predicate letters, variables, quantifiers, identity, and logical connectives. The only notion involved which could possibly be different in principle from what Kant regarded as general logic is identity, and since that is used in application to quite arbitrary objects, it does not immediately suggest a restriction as to application as the geometrical concepts do. Moreover, the schema is proved without the application of existence axioms. The range of values of the variables can be any universe whatsoever, even the empty one.[23]

Frege and his twentieth-century followers certainly thought that by their construction they had refuted the view that arithmetic depends in any way on 'pure intuition', sensibility, or time. Thus the temporal notion of the successive addition of units, or the even more concrete one of combining groups of objects, is replaced in Frege's construction by the *timeless* relation of one class being the union of two others, which can be defined in terms of the logical connective alternation as it occurs in (1). Moreover, the construction provides a framework for the application of the concept of number far beyond the scope of concrete appearances, in particular, in the elaboration of set theory.

Analogous to a non-Euclidean space would be a possible world in which the arithmetical identities turned out differently, for example, in which $2 + 2 = 5$. But would that not be a world in which there was a counter-instance to our schema, and therefore in conflict with logic? Only, of course, if the connection of meaning between '$2 + 2 = 4$' and the schema (or '$2 + 2 = 5$' and a similar schema) is preserved. I am inclined to regard the breaking of this connection as a change in the meaning of addition.

There is, however, one way out of this dilemma. With '$2 + 2 = 5$' we would associate the schema

$$(\exists_2 x)Fx \cdot (\exists_2 x)Gx \cdot (x) - (Fx \cdot Gx) \cdot \supset (\exists_5 x)(Fx \vee Gx) \qquad (4)$$

Now suppose we had a universe U in which for any choice of extensions of 'F' and 'G' this schema came out true. Even according to our notions of logic, there is a possible case in which this happens, and in which (since (1) is valid) there is also no conflict with (1), namely in which U contains fewer than four elements. In that event the antecedent of the above would always be false.[24]

If one considers the minimal existence axioms which would be needed to prove the categorical '$2 + 2 = 4$' in modern set theory, we find that again they require the universe to contain at least four elements, which can be identified with the numbers 1, 2, 3, 4.

If we accept first order quantification theory with identity as a logical framework, then it seems that we can maintain the symmetry of arithmetic and geometry in a weak sense, that such propositions as '2 + 2 = 4' imply or presuppose existence assumptions which it is logically possible to deny. To draw the line at this point and to declare thus that set theory is not logic seems to me eminently reasonable; but I shall not argue for this now, particularly since I have done so elsewhere.[25]

I think the presence of existence propositions in mathematics is one of the considerations at stake in Kant's views on mathematics, but it is not clearly differentiated from others. His general views on existence imply that existential propositions are synthetic, but he never applies this doctrine directly to the existence of abstract entites. In the letter to Schultz cited above, Kant says that arithmetic, although it does not have axioms, does have postulates. Postulates as to the possibility of certain constructions, for Kant constructions in intuition, played the role of existence assumptions in Euclidean geometry. Schultz states as a postulate in the *Prüfung* essentially that addition is defined.

This factor is also present in Kant's remarks about 'construction of concepts in pure intuition', which he regards as the distinguishing feature of mathematical method. If the geometer wants to prove that the sum of the angles of a triangle is two right angles, he begins by *constructing a triangle* (A 716/B 744). This triangle, as we indicated above, can serve as a paradigm of all triangles; although it is itself an individual triangle, nothing is used about it in the proof which is not also true of all triangles. The proof consists of a sequence of constructions and operations on the triangle.

Kant's view was that it is by this construction that the concepts involved are developed and the existence of mathematical objects falling under them is shown. Although we need not regard this theorem as implying or presupposing that there are triangles, Kant regarded a general proposition as empty, as not genuine knowledge, if there are no objects to which it applies. In this instance only the construction of a triangle can assure us of this. Apart from that, further existence assumptions are used in the course of the proof, in the example of A 716/B 744 of extensions of lines and of parallels.

The same factor is also suggested in the rather puzzling passage in which Kant says that the operation with variables, function symbols, and identity in traditional algebraic calculation involves 'exhibiting in intuition' the operations involved, which he calls 'symbolic construction'. In fact, such operation presupposes that the functions involved are *defined* for the arguments we permit ourselves to substitute for the variables. Moreover, the construction of an algebraic expression for an object to satisfy a certain condition is the very paradigm of a constructive proof of the existence of such an object. However, I think there is something else at stake in this passage, which I shall come to.

It is by no means obvious that the existence assumptions which must be made in the deductive development of mathematics have any connection with sensibility and its alleged form. Frege for one was quite convinced that they did not. What Kant says that bears on this point is not completely clear, partly because in the nature of the case it is bound up with some difficult notions in his philosophy, partly because again he did not disengage this issue from some others.

As a preliminary remark, we must observe that Kant certainly did not regard arithmetic as a special theory of, say, time, in the sense in which he regarded geometry as a special theory of space. It does not turn up in this connection in the proofs of the apriority of time in either the Aesthetic or the corresponding discussion in the Inaugural Dissertation (§ 12, §14 no. 5).

Nevertheless it is clear that according to Kant, the dependence of arithmetic on the forms of our intuition is in the first instance only on time. I should venture to say that space enters the picture only through the general manner in which inner sense, and thus time, depends on outer sense, and thus space. We shall be clear about the intuitive character of arithmetic when we are clear about the manner in which it depends upon time.

Whenever Kant speaks about this subject, he claims that number, and therefore arithmetic, involves *succession* in a crucial way. Thus in arguing that intuition is necessary to see that $7 + 5 = 12$:

For starting with the number 7, and for the concept of 5 calling in the aid of the fingers of my hand as intuition, I now add *one by one* to the number 7 the units which I previously took together to form the number 5, and with the aid of that figure [the hand] see the number 12 *come into being*. (B 15–16, emphasis mine.)

When he gives a general characterization of number in the Schematism, the reference to succession occurs essentially:

The pure image of all magnitudes (*quantorum*) for outer sense is space; that of all objects of the senses in general is time. But the pure *schema* of magnitude (*quantitatis*), as a concept of the understanding, is *number*, a representation which comprises the successive addition of homogeneous units. [A 142/B 182.]

As I said, this seems to conflict not only with the interpretation which number and addition acquire in such constructions as Frege's, in which instead of the *successive addition* of 'units' we have a *timeless* relation, for example, that one set is the union of two others; but also with the application of these notions within modern mathematics, in which arithmetical statements can be made about structures which are entirely timeless, and in reference to which any talk of 'successive addition' is on the face of it entirely metaphorical.

In the letter to Schultz, Kant qualifies his position in a way which does more justice to this more general character of arithmetic:

Time, as you quite rightly remark, has no influence on the properties of numbers (as pure determinations of magnitude), as it does on the property of any alteration (as a quantum), which itself is possible only relative to a specific condition of inner sense and its form (time); and the science of number, in spite of the succession, which every construction of magnitude [*Größe*] requires, is a pure intellectual synthesis which we represent to ourselves in our thoughts.

Earlier in the letter he writes:

Arithmetic, to be sure, has no axioms, because it actually does not have a *quantum*, i.e., an object of intuition as magnitude, for its object, but merely *quantity*, i.e., a concept of a thing in general by determination of magnitude.

Kant is here in fact reaffirming a position affirmed in the Dissertation:

To these there is added a certain concept which, though itself indeed intellectual, yet demands for its actualization in the concrete the auxiliary notions of time and space (in the successive addition and simultaneous juxtaposition of a plurality), namely, the concept of number, treated of by arithmetic. [§ 12.]

These remarks place arithmetic less on the intuitive and more on the conceptual side of our knowledge. If arithmetic had for its object 'an object of intuition as magnitude', i.e., forms such as the points, lines, and figures of geometry, then it would refer quite directly to a form of intuition. But instead it refers to 'a concept of a thing in general'; the science of number is a 'pure intellectual synthesis'. This latter phrase especially suggests that arithmetical notions might be definable in terms of the pure categories and thus be associated with logical forms which do not refer at all to conditions of sensibility. Such a view would seem to conflict with the statement of the Schematism that number is a schema.

The reference to 'a concept of a thing in general' is no doubt to be meant in the same sense as that in which the categories are said to specify the concept of an object in general, and the pure intellectual synthesis is no doubt that of the second edition transcendental deduction, which is the synthesis of a manifold of intuition in general, which is for us realized so as to yield knowledge only in application to intuitions according to *our* forms of intuition. Thus the 'concept of an object in general' could give rise to actual knowledge *of objects* only if these objects can be given according to our forms of intuition.

But does this merely mean that objects in space and time provide the only concrete application of these concepts which we can know to exist, as one might expect from the absence of special reference to intuition? Whether it means this or something more drastic is, I think, a special case of the general dilemma about the understanding which I mentioned in the beginning. In either case, however, it would be a plausible interpretation of Kant to say that the forms of intuition must be appealed to in order to verify the existence assumptions of mathematics.

However, it is not very clear how to apply the general conceptions derived from the Aesthetic and the Transcendental Deduction to the case at hand. The direct existence propositions in pure mathematics are of *abstract* entities, and it is only in the geometric case that they can be said to be in space and time. I do think that the objects considered in arithmetic and predicative set theory can be construed as *forms* of spatio-temporal objects. Full set theory would of course not be accommodated in this way, but it is not reasonable to expect that from a Kantian point of view impredicative set theory should be intuitive knowledge or indeed genuine knowledge at all. It could legitimately be said to postulate entities beyond the field of possible experience.[26]

It is natural to think of the natural numbers as represented to the senses (and of course in space and time) by numerals. This does not mean mainly that numerals function as names of numbers, although of course they do, but that they provide instances of the structure of the natural numbers. In the algebraic sense, the set of numerals generated by some procedure is isomorphic to the natural numbers in that it has an initial element (e.g., '0') and a successor relation which the notion of natural number requires. In this sense, of course, the numerals are abstract mathematical objects; they can be taken as geometric figures. But of course concrete tokens of the first n numerals are likewise a model of the numbers from 1 to n or from 0 to $n-1$. A set of objects has n elements if it can be brought into one-to-one correspondence with the numbers from 1 to n; a standard way of doing this is by bringing them in some order into correspondence with certain numerals representing these numbers, that is by counting. (The numerals used in work in formal logic, for example where the initial element is '0' and the $(n + 1)$st numeral is obtained by prefixing 'S' to the nth numeral, have the further property that each numeral contains within itself all the previous ones so that the nth numeral is itself a model of the numbers from 0 to n.)

The basis for the use of a concrete perception of a sequence of n terms in verifying general propositions is that, since it serves as a representative of a structure, the same purpose could be served by any other instance of the same structure, that is any other perceptible sequence which can be placed in a one-one correspondence with the given one so as to preserve the successor relation. This might justify us in calling such a perception a 'formal intuition.' We might note that the physical existence of the objects is not directly necessary, so that we can abstract also from that 'material' factor.

An empirical intuition functions, we might say, as a pure intuition if it is taken as a representative of an abstract structure. Such a perception provides the fullest possible realization before the mind of an abstract concept. One of the important questions about Kant's philosophy of arithmetic is whether a comparable realization exists beyond the limits of scale of concrete perception.

Before we can enter into this question, let me point out another closely related reason in Kant's mind for regarding mathematics as dependent on intuition. This comes out in particular in the concept of 'symbolic construction'. The algebraist, according to Kant, is getting results by manipulating *symbols* according to certain rules, which he would not be able to get without an analogous intuitive representation of his concepts. The 'symbolic construction' is essentially a construction with *symbols* as objects of intuition:

Once it [mathematics] has adopted a notation for the general concept of magnitudes so far as their different relations are concerned, it exhibits in intuition, in accordance with certain universal rules, all the various operations through which the magnitudes are produced and modified. When, for instance, one magnitude is to be divided by another, their symbols are placed together, in accordance with the sign for division, and similarly in the other processes; and thus in algebra by means of a symbolic construction, just as in geometry by means of an ostensive construction (the geometrical construction of the objects themselves), we succeed in arriving at results which discursive knowledge could never have reached by means of mere concepts [A 717/B 745].

That this is a source of the clarity and evidence of mathematics and provides a connection of mathematics with sensibility is indicated by the following remark:

This method, in addition to its heuristic advantages, secures all inferences against error by setting each one before our eyes [A 734/B 762].

A connection of mathematics and the senses by way of symbolic operations is already claimed in Kant's prize essay of 1764, *Untersuchung über die Deutlichkeit der Grundsätze der natürlichen Theologie und der Moral*,[27] which presents a prototype of the theory of mathematical and philosophical method of the Discipline of Pure Reason in its Dogmatic Employment. For example, consider the statement of the latter:

Thus philosophical knowledge considers the particular only in the universal, mathematical knowledge the universal in the particular, or even in the single instance, although still always *a priori* and by means of reason [A 714/B 742].

This distinction corresponds in the prize essay to the following, where the distinctive role of signs in mathematics is explicitly emphasized:

Mathematics considers in its solutions, proofs, and inferences the universal *in* [*unter*] the signs *in concreto*, philosophy the universal *through* [*durch*] the signs *in abstracto*.[28]

The certainty of mathematics is connected with the fact that the signs are *sensible*:

Since the signs of mathematics are sensible means of knowledge, one can know with the same confidence with which one is assured of what one sees with one's own eyes that one has not left any concept out of account, that every equation has been derived by easy rules, etc.; thereby attention is made much easier in that it must take account only of the signs as they are known individually, not the things as they are represented generally.[29]

The prize essay suggests a position incompatible with the *Critique of Pure Reason*, namely that since in mathematics signs are manipulated according to rules which we have laid down (in contrast to philosophy, where the value of any definition turns on its having a certain degree of faithfulness to pre-analytic usage), operation with signs according to the rules, without attention to what they signify, is in itself a sufficient guarantee of correctness.[30]

These passages show that a connection between *sensibility* and the intuitive character of mathematics existed in Kant's mind before he developed the theory of space and time of the Aesthetic. However, unlike in the later work, no inference is drawn at this stage from this connection to a limitation of the application of mathematics to sensible *objects*.

The general point behind the observations on symbolic construction can be put in the following way: in general, a mathematical proposition can be verified only on the basis of a proof or calculation, which is itself a construction in intuition. But in view of the remarks about '$7 + 5 = 12$', a more special fact may have influenced Kant. Certain 'symbolic constructions' associated with propositions about number actually involve constructions isomorphic to the numbers themselves and their relations, or at least an aspect of them. Thus in Leibniz's proof that $2 + 2 = 4$, '$2 + 2$' must be written out as '$2 + (1 + 1)$' and the two 1s as it were added on to the '2'. A corresponding proof of '$7 + 5 = 12$' would involve *five* such steps instead of two.

A similar observation concerning the schema (1) has been made by a number of writers. Although the schema does not *imply* that the universe contains any elements or that any construction can be carried out, the *proof* of it involves writing down a group of two symbols representing the *F*s, another such group representing the *G*s, and putting them together to get four symbols. So that it is not at all clear that '$2 + 2 = 4$' interpreted as a proposition about the combinations of symbols is not more elementary than the logically valid schema (1).

I have already suggested that the 'symbolic' construction in generating numerals is already enough to settle the question of their references. In the same way the actual carrying out of the calculations shows the well-defined character for individual arguments of recursively defined functions. However, induction, which I have wanted to leave out of account here, is involved in seeing that they are defined for all arguments. Maybe Kant ought to have said that apart from intuition I do not even know that there is such a number as '7 + 5'. And it seems that one could not see by a particular construction that there is such a number without also seeing it to be 12. This is in agreement with Hintikka's statement that the sense of Kant's statement that numerical formulae are indemonstrable is that the construction required for their proof is already sufficient.

The considerations about the role of symbolic operations apply equally to logic and therefore undermine Kant's apparent wish to distinguish them on this basis. This appears more forcefully in modern logic, where instead of a short list of forms of valid inference one has an infinite list which must be specified by some inductive condition. In my opinion this is a consequence to be accepted and is even in general accord with Kant's statements that synthesis underlies even the possibility of analytic judgements.

The special connection of arithmetic and time can, I think, be explained as follows: if one constructs in some way, such as on paper or in one's head, such a sequence of symbols as the first n numerals, the structure is already represented in the sequence of operations and more generally in the succession of mental acts of running through a group of n objects, as in counting. Thus time enters in through the succession of acts involved in construction or in successive apprehension. This connects with Kant's remark about number in the Schematism. In the operations involved in representing a number to the senses, we also generate a structure in *time* which represents the number. Time provides a universal source of models for the numbers. In particular, Kant held that it is only by way of successively perceiving different aspects of a manifold and yet keeping them in mind as aspects of one intuition that we can have a clear conception of a plurality. For quite small numbers this seems doubtful although not for larger ones. Nonetheless the element of succession appears even for the smaller ones in the comparison involved in generating or perceiving them in *order*, and the order is certainly part of our concept of number. What would give time a special role in our concept of *number* which it does not have in general is not its necessity, since time is in some way or other necessary for all concepts, nor an explicit reference to time in numerical statements, which does not exist, but its sufficiency, because the temporal order provides a representative of the number which is present to our consciousness if any is present at all.

Of course it is one thing to speak of representation in space and time and another to speak of representation to *the senses*. What is represented to the senses is presumably represented in space and time, but maybe not vice versa. To establish a link of these two Kant would appeal to his theory of space and time as forms of sensibility. The relevant part of this theory is that the structures which can be represented in space and time are structures of *possible* objects of perception. The kind of possibility at stake here must be essentially mathematical and go beyond 'practical' or physical possibility.[31]

Consider once again a procedure for generating numerals, say by starting with '0' and prefixing occurrences of '*S*'. The actual use of these as symbols requires that they be perceptible objects. Nonetheless we say it is *possible* to iterate the procedure indefinitely and therefore to construct indefinitely many numerals. Thus it is clear that the numerals (numeral types)[32] which it is in this sense possible to construct extend far beyond the numeral-tokens which have ever been produced in history or which could in any concrete sense actually be used as symbols.[33] This possibility of iteration is necessary for the constructibility of indefinitely many numerals and therefore for the infinity of natural numbers to be given by intuitive construction. Moreover, some insight into such iteration seems necessary for mathematical induction.

Insofar as the appeal to pure intuition for the evidence of mathematical statements is supposed to be an analogy of mathematical and perceptual knowledge, it holds less well for propositions involving the concept of indefinite iteration, such as these proved by induction, than for propositions such as $2 + 2 = 4$. There seem to be two independent types of insight into our forms of intuition which a Kantian view requires us to have, that which allows a particular perception to function as a 'formal intuition' and that which we have into the *possible progression* of the generation of intuitions according to a rule. To speak of a peculiar *kind* of intuition in the second case seems quite misleading. The mathematical knowledge involved has a highly complex relation to 'intuition' in the more specifically Kantian case.

That complexity must be in some way present in the 'intuitions' of space and time since space is an individual which is *given*, but its structure also determines the limits of *possible* experience and contains various infinite aspects. No doubt the plausibility of the idea that space is present in immediate experience made it more difficult for Kant to appreciate the differences of the kinds of evidences covered by his notion of pure intuition. I am sure that more could be done to explicate the Kantian view of their connection.

In our discussion of intuition, we have somewhat lost sight of the view of logic which at the start we attributed to Kant, which except for the question of existence resembles the modern views called Platonist. Although Kant's view of intuition fits better with the modern tendencies called constructivist or intuitionist, it seems certain that the concept of pure intuition was meant to go with this view of logic and not to replace it. Without using notions like 'concept' and 'object' in a quite general way, it is probably not possible to describe it. It would be hasty for that reason to identify Kant's conception of intuition with that of Dutch intuitionists, although Brouwer's undoubtedly shows some affinity. It would also be hasty to regard Brouwer's critique of classical mathematics as altogether in accord with Kantianism.

[1] An earlier version of this paper was written while the author was George Santayana Fellow in Philosophy, Harvard University, and presented in lectures in 1964 to the University of Amsterdam and the Netherlands Society for Logic and the Philosophy of Science. I am indebted to J. J. de Iongh, J. F. Staal, and G. A. van der Wal for helpful comments. I am also grateful to Jaakko Hintikka for sending me two unpublished papers on the subject of this paper.

[2] All passages are quoted in the translation of Norman Kemp Smith (London: Macmillan, 1929) with slight modifications. Other translations from German are my own. Translations of Kant's Inaugural Dissertation are by John Handyside, in *Kant's Inaugural Dissertation and Early Writings on Space* (Chicago and London: Open Court, 1929).

[3] *Kants Gesammelte Schriften*, ed. by the Prussian Academy of Sciences (Berlin, 1902–), IX 91. This edition will be referred to as 'Ak'.

[4] 'Kant's "new method of thought" and his theory of mathematics', *Ajatus* 27 (1965), p. 43. Hintikka argues in detail for this thesis in 'On Kant's notion of intuition (*Anschauung*)', in T. Penelhum and J. J. MacIntosh, eds., *The First Critique* (Belmont, Calif.: Wadsworth, 1969). The same idea seems to underlie the analysis of Kant's theory of mathematical proof in E. W. Beth, 'Über Lockes "allgemeines Dreieck" ' in *Kant-Studien* 48 (1956–7), 361–80.

[5] 'It is a mere tautology to speak of general or common concepts'. (*Logic*, para. 1, Ak. IX 91.)

[6] One might attribute to Kant the view that there are no such representations. The classification Kant makes in A 320/B 376–7 and *Logic* § 1 is of *Erkenntnisse*, which Kemp Smith translates as 'modes of knowledge' but which in many contexts would be more accurately though inelegantly translated as 'pieces of knowledge'. Then the relation of a representation to its object is that through which one can *know* its object, and it might be held that intuition in the full sense is the only singular representation which can provide such knowledge. This view would have the perhaps embarrassing consequence that an object which is not in some way perceived is not really known as an individual.

[7] Cf. the examples of 'truths of reason' given by Leibniz, *Nouveaux Essais*, IV. ii. 1.

[8] Ibid. IV. vii. 10.

[9] *Arithmetik und Kombinatorik bei Kant* (Diss, Freiburg, 1934) (Itzehoe, 1939); *Kant's Metaphysics and Theory of Science* (Manchester University Press, 1953), ch. 1; *Klassische Ontologie der Zahl, Kant-Studien* Ergänzungsheft 70 (1956), sect. 12.

[10] Neither Leibniz nor Schultz seems to mention the fact that in order to prove formulae involving *multiplication*, such as '$2 \cdot 3 = 6$', one also needs instances of the distributive law.

[11] '1. Aus mehreren gegebenen gleichartigen Quantis *durch ihre successive Verknüpfung* den Begriff von einem Quanto zu erzeugen, d.i. sie *in ein Ganzes* zu verwandeln. 2. Ein jedes gegebenes Quantum, *um so viel, als man will*, d.i. *sie ins Unendliche zu vergrößern, und zu vermindern.*' (*Prüfung*, I, 221.)

[12] Ak. X 554–558.

[13] *Arithmetik und Kombinatorik bei Kant*, p. 57.

[14] C. J. Gerhardt (ed.), *Leibnizens mathematische Schriften* (Halle: Schmidt, 1849–1863), VII 78. Leibniz gives a definition of addition from which he claims commutativity follows immediately. One could read his argument as deriving the commutativity of addition from the commutativity of set-theoretic union.

[15] 'Über Lockes "allgemeines Dreieck".'

[16] 'Kant's "new method of thought" '. 'On Kant's concept of intuition', also 'Are logical truths analytic?' *Philosophical Review*, **74** (1965) 178–203, 'Kant on the mathematical method', *The Monist* **51** (1967), 352–75.

[17] It ought to be remarked that while no doubt the distinction which Kant makes between axioms and postulates derives historically from that of 'common notions' and postulates in Euclid, Kant's distinction does not correspond exactly to Euclid's. Euclid's division is between more general principles and specifically geometrical ones. For Kant postulates are 'immediately certain practical judgements,' the action involved is construction, and their purport is that a construction of a certain kind can be carried out. The role they play is thus that of existence axioms. Euclid's common notions are all of a type which Kant asserted to be analytic propositions (A 164/B 204, B 17), while axioms proper must be synthetic.

[18] Op. cit. p. 365.

[19] Cf. W. V. Quine, *Methods of Logic* (New York: Holt, Rinehart and Winston, revised ed. 1959), sect. 28.

[20] 'Kant's "new method of thought" ', p. 43, also 'Kant on the mathematical method'. The texts are A 717/B 745, A 734/B 762.

[21] In 'Are logical truths analytic?' Hintikka develops a distinction between analytic and synthetic according to which some logical truths are synthetic. He suggests that the logical truths which are analytic according to this criterion are roughly those which Kant would have regarded as analytic. It follows, however, that in some of the arguments which according to Beth and Hintikka involve for Kant an appeal to intuition, the conditional of their premises and conclusion is analytic. In particular, this is true of the example that Beth works out in detail in 'Über Lockes "allgemeines Dreieck" ', § 7. In order to be applied to mathematical examples like Kant's, Hintikka's criterion would have to be extended to languages containing function symbols. The way of doing this which seems to me most in the spirit of Hintikka's definition has some anomalous consequences.

See also 'An analysis of analyticity', 'Are logical truths tautologies?', 'Kant vindicated', and 'Kant and the tradition of analysis', in P. Weingartner (ed.), *Deskription, Analytizität, und Existenz* (Salzburg and Munich: Pustet, 1966).

[22] Beth, 363.

[23] In fact, (1) is analytic according to the criterion of 'Are logical truths analytic?' (see note 21 above). However, according to another criterion which might be more in the spirit of Kant, to consider as synthetic a conditional whose proof involves formulae of degree higher than its *antecedent*, (1) is synthetic. Hintikka takes account of this in 'Are logical truths analytic?' by making an additional distinction between analytic and synthetic *arguments*, such that in the relevant sense the argument from the conjuncts of the antecedent of (1) as premises to its consequent as conclusion is synthetic.

[24] Cf. Hao Wang, 'Process and existence in mathematics', Y. Bar-Hillel, E. I. J. Poznanski, M. O. Rabin, A. Robinson (eds.), *Essays in the Foundations of Mathematics*, dedicated to A. A. Fraenkel (Jerusalem: Magnes Press, 1961), p. 335.

[25] 'Frege's theory of number', Max Black (ed.), *Philosophy in America* (London: Allen & Unwin, 1965), 180–203.

[26] An interesting intermediate case is how constructive proofs as the object of intuitionist mathematics could be interpreted from a Kantian point of view. According to Kant as I interpret him, certain empirical constructions can function as paradigms so as to establish necessary truths because of the intention or meaning associated with them. Intuitionism would require that our insight into these meanings be sufficient not only to establish laws directly relating to objects in space and time but also to establish laws concerning the *intentions* as 'mental constructions'. I leave open the question of whether this is possible from Kant's point of view or not.

[27] Ak. II 272–301. It is translated by G. B. Kerferd and D. E. Walford in *Kant: Selected Pre-Critical Writings* (Manchester University Press, 1968). In this volume, which also includes a translation of the Inaugural Dissertation, the Ak. pagination is given in the margins.

[28] 'Die Mathematik betrachtet in ihren Auflösungen, Beweisen, und Folgerungen das Allgemeine unter den Zeichen *in concreto*, die Weltweisheit das Allgemeine durch die Zeichen *in abstracto*.' (Erste Betrachtung, § 2, heading, Ak. II 278).

[29] 'Denn da die Zeichen der Mathematik sinnliche Erkenntnismittel sind, so kann man mit derselben Zuversicht, wie man dessen, was man mit Augen sieht, versichert ist, auch wissen, daß man keinen Begriff aus der Acht gelassen, daß eine jede einzelne Vergleichung nach leichten Regeln geschehen sei u.s.w. Wobei die Aufmerksamkeit dadurch sehr erleichtert wird, daß sie nicht die Sachen in ihrer allgemeinen Vorstellung, sondern die Zeichen in ihrer einzelnen Erkenntnis, die da sinnlich ist, zu gedenken hat.' (Dritte Betrachtung. § 1, Ak. II 291).

[30] But cf. the following: 'In der Geometrie, wo die zeichen mit den bezeichneten Sachen überdem eine Ähnlichkeit haben, ist daher diese Evidenz noch größer, obgleich in der Buchstabenrechnung die Gewißheit eben so zuverlässig ist.' (Ibid., 292).

[31] One might say that it is possible to construct tokens. The sense of possibility in which this is possible is, however, derivative from the mathematical possibility of constructing types (or mathematical *existence* of the types). For we declare that the tokens are possible either directly on the basis of the mathematical construction, or physically on the basis of a theory in which a mathematical space which is in some way infinite is an ingredient.

[32] Cf. my 'Infinity and Kant's conception of the "possibility of experience" ', *Philosophical Review* **73** (1964), 182–97.

[33] This does not imply that there is an upper limit on the numbers which can be individually represented, once we admit notations for faster-growing functions than the successor function. This happens already in Arabic numeral notation. The number 1,000,000,000,000, if written in '0' and 'S' notation with four symbols per centimeter, would extend from the earth to the moon. That there is such an upper limit follows, of course, from the assumption that human history must come to an end after a finite time.

II

VISUAL GEOMETRY

JAMES HOPKINS

I

KANT thought Euclid's geometry true of everything spatially intuitable.
This implied that only Euclid's geometry – Euclidean figures or a Euclidean
space – could be seen, imagined, or visualized. To many, including modern
philosophers, this has seemed true. Thus Frege:

Empirical propositions hold good of what is physically or psychologically actual, the
truths of [Euclidean] geometry govern all that is spatially intuitable, whether actual
or product of our fancy. The wildest visions of delirium, the boldest inventions of
legend and poetry . . . all these remain, so long as they remain intuitable, still subject
to the axioms of geometry. Conceptual thought can after a fashion shake off this
yoke, when it assumes, say, a space . . . of positive curvature. To study such
conceptions is not useless by any means; but it is to leave the ground of intuition
entirely behind. If we do make use of intuition even here, as an aid, it is still the same
old intuition of Euclidean space, the only space of which we have any picture.[1]

Bennett:

If we restricted ourselves to what could be 'imagined' or seen at a glance, then
perhaps we should be bound to regard spaces as Euclidean . . . it is not clear how one
could see at a glance that two straight lines intersected twice: it seems that if both
intersections are seen at once then at least one of the lines must look curved.[2]

And Strawson:

Consider the proposition that not more than one straight line can be drawn between
any two points. The natural way to satisfy ourselves of the truth of this axiom of
phenomenal geometry is to consider an actual or imagined figure. When we do this, it
becomes evident that we cannot, either in imagination or on paper, give ourselves a
picture such that we are prepared to say of it both that it shows two distinct straight
lines and that it shows both these lines as drawn through the same two points.[3].

Surely what Bennett and Strawson say here is true. We cannot see or
picture two definitely straight lines between two points. Given two points
we can picture one definitely straight line between them; but any other we
picture will be curved. For example:

From *Philosophical Review* **82** (1973), pp. 3–34. Reprinted by permission of the author and the
Philosophical Review.

So it seems we can form the Euclidean but not the non-Euclidean picture. Similarly in other cases: we picture triangles of the same shape but different sizes, whose angles equal two right angles. Such are Euclid's, and we can imagine no others. So our pictures do suggest, as Frege believed, that we see, imagine, or picture anything whatever as Euclidean – as spatially disposed, and hence geometrically describable, in no other than Euclid's terms.

This belief is part of the content of the Kantian theory that the form of outer sense is Euclidean. Also it was a source of Kant's conviction that Euclid's propositions were known true *a priori*. Kant assumed that geometric proof required construction on a figure. He thought the proved propositions synthetic because this construction was a synthesis to be contrasted with the analysis of concepts. He thought them known *a priori* because the construction was not taken from experience. And the construction, the picture, inevitably yielded Euclidean results.

[M]athematical knowledge . . . is the knowledge gained by reason from the construction of concepts . . . I construct a triangle by representing the object which corresponds to the concept either in the imagination alone, in pure intuition, or in accordance therewith also on paper, in empirical intuition – in both cases completely *a priori*, without having borrowed the pattern from any experience.[4]

Frege knew of modern developments in geometry and had more sophisticated reasons for regarding Euclidean propositions as synthetic. Yet he seems to have thought them known *a priori*, solely because they alone could be intuited. 'In calling the truths of geometry synthetic and *a priori* [Kant] revealed their true nature.'[5]

Now it is commonplace that Kant's beliefs about geometry have been superseded. The refinement of geometry as an abstract science has made clear that construction on a figure has no such role in proof as Kant supposed. The discovery of non-Euclidean geometries has been taken to show that the truth or falsity of Euclid's description of space is an empirical matter. And it is widely accepted that the successful use of a non-Euclidean geometry in Einstein's theory of relativity – in which, for example, there may be two straight lines (two paths as short as, or shorter than, any other) between two points – has established that Euclidean geometry is as a matter of fact false of physical space.

Euclidean and non-Euclidean geometries contain inconsistent statements about straight lines. So on a consistent interpretation of 'straight line' only one geometry can be true. On interpretations which are common, plausible, and scientifically useful, the geometry of space according to Einstein's theory is not in general Euclidean. Einstein writes:

Euclidean geometry does not hold even to a first approximation in the gravitational field, if we wish to take one and the same rod, independently of its place and orientation, as a realization of the same interval.[6]

And Barker describes the situation as follows:

Suppose a closed three-sided figure is laid out, its sides being determined by light rays, or by paths along which measuring rods need be laid down the fewest times, or by paths along which stretched cords tend to lie. According to the theory of relativity, we must predict that in the presence of a gravitational field the sum of the angles of this figure will be greater than two right angles. We must also predict that between any two separate points there will, in the presence of gravitational fields, be more than one path along which light rays can travel, more than one path divisible into overlapping sub-segments along each of which a measuring rod need be laid down a minimum number of times to get from end point to end point, and so forth.

Many . . . would say that the theory of relativity proves space to be Riemannian rather than Euclidean in its general form. Einstein himself is the outstanding representative of this viewpoint, which he expressed in more generalized form in his often-quoted dictum 'As far as the laws of mathematics refer to reality, they are not certain; and as far as they are certain, they do not refer to reality.'[7]

But here, surely, is a problem. It seems that science has given reason for believing that Euclidean geometry is false, that physical space may most accurately be described by a non-Euclidean geometry. Yet examples lead us to suppose that the only space we can imagine, picture, or visualize, is one described by Euclidean geometry. But the space it seems we must picture as Euclidean is the same space as that which, on scientific grounds, is judged non-Euclidean. And why, one might ask, can we not picture our space as science gives reason to believe it is? How are we constrained to see, imagine, or visualize it in terms of a theory inconsistent with what we might believe true of it on scientific grounds?

II

It may be thought relevant that there are familiar ways of representing a non-Euclidean space. Popular scientists and mathematicians sometimes draw gently curving arcs, which may intersect twice, to represent non-Euclidean straight lines. Again, the surface of a sphere provides a model for a Riemannian (non-Euclidean) space. An arc of a great circle is the shortest path between two points on the surface of a sphere; and many propositions about lines and figures in a Riemannian space hold for great circles and spherical figures. Thus there may be two great circle paths between (antipodal) points, the sum of the angles of a closed figure bounded by three great circles is more than two right angles, and so forth.

Diagrams and models of this kind are often invoked in connection with the problems of picturing non-Euclidean space. But clearly they cannot help us picture space as non-Euclidean. For arcs on paper and great circles on a sphere both are, and are seen or pictured as, curved lines. So in using such a diagram or model, we picture curved lines, but not straight lines, intersecting twice. And it gets us no further to try to picture space as *in accord with* the diagram, or *on the model* of the sphere. In failing to picture distinct straight lines intersecting twice, we fail to imagine straight lines with the relevant characteristics of the diagram or model. So we thereby fail to picture in accord with the diagram or on the model of a sphere.

There is a deeper and more sophisticated approach to visualizing the non-Euclidean. Philosophers and mathematicians, among them Reichenbach,[8] describe visibly non-Euclidean worlds. Their descriptions are meant to enable us to form imaginative pictures of the worlds.

Since in either Euclidean or non-Euclidean geometry a straight line is the shortest path between two points, it is clear that which lines in a given manifold are straight will be determined by measurement. So it is possible, for example, to describe a world in which we see two paths between two points and find both to be straight, say by measuring with rods rigid by every test to show both equal and any alternative longer.

With descriptions of this kind in mind, persons often claim to form non-Euclidean pictures. But on investigation it seems we cannot really do so. Consider, for example, lines such as *a* and *b*.

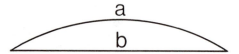

We have an interpretation, consistent with Euclid, of how *a* and *b* look. On the face of it, and without any special story in mind, we see or picture *a* as curved, *b* as (approximately) straight. Now suppose we try to imagine a world in which both are straight – that is, equal and shorter than any alternative between their intersections.

We can, for example, picture *a* and *b* as being measured to give this result. But this is not yet imagining them straight. For it is no different from imagining rods to grow in measuring *a* or shrink in measuring *b* to give the result. So imagining *a* equal to *b* is not yet distinguished from imagining *a* curved but measured equal to *b* by unstable instruments. As regards the length or straightness of *a* and *b* we have imagined nothing new. Our picture, our way of seeing their length or straightness, has not changed. So no different description is justified.

It adds nothing to *speak* of the rods, or the unit of measure, as staying the same in measuring *a* and *b* to show them equal. What is in question is whether we actually picture this. Even if we add other pictures, and imagine rods rigid by other tests in other contexts, we are still required to picture them rigid in measuring *a* and *b* equal. As long as our picture of *a* and *b* does not change, we cannot do so.

So it seems no matter what story we tell, or what additional pictures we form, we do not picture anything different from that with which we began – namely, *a* curved and *b* straight. So we do not picture a situation different from the Euclidean. Thus even by means of this more sophisticated approach, we do not succeed in picturing a non-Euclidean situation.

A comparison may make this clearer. There are cases in which two ways of seeing, between which we can change, are available. We can picture other lines on the paper – a duck–rabbit, say – first one way (as a duck), then another (as a rabbit). Here the way of seeing does change, and there is an experience of change of sight. The change can be induced by giving descriptions by interpolating other pictures. With our way of seeing geometrically it is not like this. We cannot first see *a* as longer than *b*, then see them as equal. Here no change occurs and no pictures or descriptions induce one.

Reichenbach calls such a change in the way of picturing lengths as would constitute non-Euclidean picturing an adjustment in congruence or an emancipation from Euclidean congruence. Thus he says in a parallel case that during the adjustment 'one can forget that from the viewpoint of Euclidean geometry these distances are different in length.' (p. 56.) But we cannot forget that *a* and *b* are different in length and see them as equal. This shows, in Reichenbach's terms, that no adjustment or emancipation from Euclidean congruence takes place. And Reichenbach says 'so long as we cannot emancipate ourselves from Euclidean congruence . . . non-Euclidean relations can only be mapped upon the visualized Euclidean space.' (p. 57.) It follows that our visualized space remains Euclidean.

So we are left with the difficulty of picturing space as other than Euclidean. This may explain what Frege meant, when he said of the study of non-Euclidean geometry:

If we do make use of intuition even here, as an aid, it is still the same old intuition of Euclidean space, the only space of which we have any picture.

for, as he continues,

only then the intuition is not taken at its face value, but as symbolic of something else; for example we call straight or plane what we actually intuit as curved.[9]

III

This difficulty, among others, is sharply presented by the final section of *The Bounds of Sense*. Strawson there espouses a carefully qualified analogue of Kant's theory of geometry. He introduces the notion of a phenomenal figure as a correlative of Kant's object in pure intuition.

Kant said it did not matter whether 'construction of a (spatial) concept in pure intuition' took place with the aid of a figure drawn on paper *or simply in the imagination*. Now the visual imagination cannot supply us with physical figures. But it can supply us with what, for want of a better word, I will call *phenomenal* figures. . . . The straight lines which are the objects of pure intuition are not physical straight lines. They are, perhaps, phenomenal straight lines. They are not physical objects, or physical edges, which, when we see them, look straight. They are rather just the looks themselves which physical things have when, and in so far as, they look straight. An arrangement of physical lines or edges may look triangular. But it is not the physical lines, so arranged, which constitute the triangle which is the object of pure intuition; it is, rather, the triangular look which they have, the phenomenal triangle which they present. [p. 282.]

This suggests that '*X* is a phenomenal straight line (triangle, etc.)' is to be read as '*X* is the look a thing has if it looks a straight line.' A similar account can be given for Strawson's other phenomenal items, such as visual images (pp. 282, 287) and objects of sight, described as seen (the picture we give ourselves when we draw a geometric diagram).

These are distinct items – visual images, for example, are not the appearance of things – and their assimilation has caused confusion in the philosophy of perception. Here the assimilation serves a likeness. In a large range of cases we should apply or withhold many of the same predicates, among them descriptions of shape and colour, in describing the appearance of *X*; *X*, as it appears, is seen, imagined, or visualized; the visual image had in seeing or the image had in remembering, imagining, or visualizing *X*; and so forth. This links the disparate items brought under the concept of a phenomenal object, and gives the examples under discussion their suggestion of scope.

Strawson wishes to use phenomenal figures, so far as possible, to provide an interpretation for geometry. This essentially requires that for 'straight line' in a geometry we read 'phenomenal straight line', and so forth, so far as possible. On this interpretation a geometry is true so far as statements in it concerning straight lines are true of phenomenal straight lines and so forth; and this will be determinable when the corresponding facts about the way things appear, and so forth, are.

Strawson shows Euclidean statements true of phenomenal figures by the exercise in visualization similar to Bennett's already quoted. He concludes:

It seems that Euclidean geometry may also be interpreted as a body of unfalsifiable propositions about phenomenal straight lines, triangles, circles, etc.; as a body of *a priori* propositions about spatial appearances of these kinds and hence, of course, as a theory whose application is restricted to such appearances.[p. 286.]

Only phenomenal geometry, he stresses, is necessarily Euclidean. Physical geometry is not:

According to modern physics, the possibility that the structure of space is non-Euclidean is at least something more than a bare possibility . . . it appears that the findings of astro-physics are more easily accommodated by the use of a geometrical theory of space incompatible with the Euclidean. [p. 280.]

Thus according to Strawson one geometry is true of physical space while another, inconsistent with it, is true of phenomenal space, of the way things look or are seen spatially. Indeed he criticizes Kant for not providing for such a possibility.

Kant's fundamental error . . . lay in not distinguishing between Euclidean geometry in its phenomenal interpretation and Euclidean geometry in its physical interpretations. . . . Because he did not make this distinction, he supposed that the necessity which truly belongs to Euclidean geometry in its phenomenal interpretation also belongs to it in its physical interpretation. He thought that the geometry of physical space *had* to be identical with the geometry of phenomenal space. [p. 285.]

In contrast, on Strawson's account, physical and phenomenal geometry need not be the same. One is the apparent geometry of spatial things or the geometry of their appearances; the other is the geometry of spatial things themselves. For the two to differ, then, is for things to appear systematically to have one geometry while in fact having another. This is the possibility it was Kant's fundamental error to overlook.

But again, such a possibility is surely a strange one. Surely if it were so, if there were such a systematic contradiction between spatial appearance and spatial reality, it would require explanation. How could the form of outer sense differ from the form of outer things?

IV

The source of difficulty is the assertion that we should apply Euclidean predicates and withhold non-Euclidean ones in describing phenomenal figures. Philosophical writers have suggested, alternatively, that this assertion is false, or that it is true, but so explicable in familiar and un-Kantian terms as to remove the difficulty.

Lucas writes:

Ewing (1938) and Strawson (1966) have attempted to save Kant's account of geometry by maintaining that it is *a priori* true at least of phenomenal geometry – the geometry of our visual experience – that it is Euclidean. But this is just what the geometry of appearances is not. Let the reader look up at the four corners of the ceiling of his room, and judge what the apparent angle at each corner is; that is, at what angle the two lines where the walls meet the ceiling appear to him to intersect each other. If the reader imagines himself sketching each corner in turn, he will soon convince himself that all the angles are more than right angles, some considerably so. And yet the ceiling appears to be a quadrilateral. From which it would seem that the geometry of appearances is non-Euclidean, with the angles of a quadrilateral adding up to more than 360°. And so it is. [10]

The impression that phenomenal geometry is Euclidean should not be yielded so easily. Really the case is more naturally described in Strawson's (Euclidean) than Lucas's (non-Euclidean) terms. We can, as Lucas implies, describe an image or appearance by means of a sketch or two-dimensional projection. [11] In sketching (the appearance of) the corners of his ceiling, the reader may draw, produce on paper, angles visibly greater than right angles; and in this sense he can say the corners appear obtuse angles. In sketching (the appearance of) his ceiling, the reader may draw a quadrilateral; and in this sense he can say his ceiling appears quadrilateral. But in no sketch will the reader draw a quadrilateral with four visibly obtuse angles. None can be drawn. So the reader will not, in this sense, describe his ceiling, or anything else, as having the appearance of a quadrilateral whose angles add up to more than 360 degrees. This description might be got only by mixing different descriptions of different projections. Rather it seems that the reader who goes by projections will say his ceiling appears a quadrilateral of 360 degrees. The most natural description of any diagram or projection will be Euclidean; the lines will not be found to deviate from Euclid's description.

The natural account for descriptions given by means of drawings or projections holds more generally. The notion of looking X, for example, is fundamentally connected with that of looking *to be* X, so that how a thing looks is connected with how it might be judged to be. Where whether something is X can be told by looking, 'looks X' often simply means either 'looks to be X' or 'looks like a thing which looks to be X, in ways relevant to judging whether it is X'. The latter holds trivially with the former; and often it can be expanded to 'look as things which are X typically do, in ways relevant to judging whether they are X'. Such appearance-world relating uses of 'looks', 'appears', and so forth, are complex and varied. But it seems that in none of them should we say that something looked non-Euclidean. We do not know that it would be for something to look to be non-Euclidean, or to look so as to lead us to judge it non-Euclidean, or to look as non-

Euclidean things typically do, in ways relevant to judging them non-Euclidean.

Nagel, by contrast, does allow that 'we find we cannot form our images except in conformity with the Euclidean axioms'. He thinks this explicable in familiar terms.

When we perform experiments in imagination upon straight lines, in what manner are these lines envisaged? We cannot employ any arbitrary images of lines in the experiment. We must *construct* our images in a certain manner. However, if we examine the mode of construction in those cases in which we allegedly intuit the imagined figures as Euclidean, we soon notice that the Euclidean assumptions are tacitly being used as the *rules of construction*. For example, we can certainly imagine two distinct lines with two points in common. But such lines do not count as straight lines, simply because they do not satisfy the Euclidean requirements of straightness, so that we seek to form our images so as to satisfy those requirements. . . . Accordingly, if the Euclidean postulates serve as rules for constructing our mental experiments, it is not at all surprising that the experiments inevitably conform to the rules. In short, if Euclidean axioms are used as implicit definitions, they are indeed *a priori* and necessary because they then specify what sorts of things are to be counted as their own instances.[12]

The suggestion is that our images satisfy Euclidean axioms because we use the axioms as implicit definitions of the characteristics of the items we construct, and hence as rules to which their construction conforms. As Reichenbach says 'the images by which we visualize geometry are always so adjusted as to correspond to the laws we read from them'. (p. 44.) This must mean, for example, that we refuse to count imagined lines as straight, solely because they violate Euclid's axioms of straightness. This suggestion is unfounded. Consider Nagel's example:

we can certainly imagine two distinct lines with two points in common. But such lines do not count as straight lines, simply because they do not satisfy the Euclidean requirements of straightness.[13]

If we do draw or imagine such lines, we can judge both curved, or one straight and the other curved. These judgements do accord with the Euclidean axiom that straight lines do not intersect twice. They cannot, however, be based solely on the axiom. The axiom alone entails only that not both lines are straight; it could not yield the judgement that both were curved, or that one was straight or another curved. Clearly we can make such judgements. We therefore judge in accord with some criterion other than the Euclidean axiom. The criterion seems obvious: length. When we judge that both lines are curved, say, we judge that both are longer than some other which might be produced; in other cases we judge that only one is. In general, so far as visual judgements (including those we have discussed) are concerned, one criterion of straightness suffices. We can say a line between two points is straight if it is a line than which no other is shorter. This criterion of straightness is common to both Euclidean and non-Euclidean geometries. So its use does not constitute the use of Euclid's axioms as implicit definitions. The fact that our images satisfy Euclid's axioms cannot, therefore, be explained by saying that in forming images we use the axioms as implicit definitions.[14] Some other explanation is required.

V

Strawson also attempts to explicate the Euclideanness of phenomenal geometry as *a priori*; and in this he is involved in an unconventional account of some geometric propositions.

He outlines what he calls 'the positivist account' of geometry, developed to clarify the role of analytic and empirical propositions in geometry to oppose the Kantian suggestion that geometric propositions are both synthetic and *a priori*. The formalization of a geometry shows the theorems deducible from the axioms on the strength of their logical expressions alone; so the conditionals connecting axioms and theorems can be regarded as logical, necessary, or *a priori* propositions, but not as empirical or synthetic. The axioms and theorems may be taken as uninterpreted, and hence not determinate propositions; or they may be regarded as, for example, physically interpreted, and hence as empirical or synthetic propositions, but not necessary or *a priori*. (Strawson mentions in passing 'a variant of the positivist view', in which 'we do indeed secure the necessity of our axioms and theorems; but only by qualifying our announced physical interpretation of the non-logical expressions of the theory by the rule that nothing whatever is to count as a falsification of the axioms or theorems'. This is similar to the approach Nagel describes as using the axioms as implicit definitions.) Thus, according to the positivists, in geometry there are necessary (logical, *a priori*) propositions and there are empirical (synthetic) ones, but none which can be regarded as both.

Now Strawson says that the positivist account is 'in a sense true'. But he believes also that a consideration of phenomenal geometry shows it to be in a sense inadequate. He summarizes his outline:

The positivist view offers us two ways of looking at the propositions of Euclidean geometry: as formulae in an uninterpreted calculus; or as the body of logically connected empirical propositions which result from the adoption of a physical interpretation for the fundamental expressions of the formulae [p. 285.]

and to it opposes his description of phenomenal geometry:

What we have had to notice is that there is a third way, different from either of these, which is also possible, and which the positivist view neglects. . . . Euclidean geometry may also be interpreted as a body of unfalsifiable propositions about phenomenal straight lines, triangles, circles, etc.; as a body of *a priori* propositions about spatial appearances of these kinds and hence, of course, as a theory whose application is restricted to such appearances [p. 286.]

Two connected features here distinguish Strawson's third way of regarding geometry from the positivist view. First, the geometrical terms in the axiom are given a phenomenal, as opposed to a physical, interpretation. Second, the phenomenally interpreted axioms are unfalsifiable, *a priori* propositions, rather than falsifiable, empirical ones. Strawson emphasizes this latter feature. Elsewhere he speaks of the 'phenomenally analytic', and of Kant's proper recognition of 'the necessity which truly belongs to Euclidean geometry in its phenomenal interpretation'.

It is unclear precisely how, or how far, Strawson takes these features as marking off a genuinely distinct and neglected way of regarding geometry. On the surface they do not. The positivists neglected neither phenomenal geometry nor the possibility that its propositions were *a priori*. Reichenbach wrote at length about geometric visualization; and he and Nagel, as we saw, describe a phenomenal interpretation of Euclidean geometry as true *a priori* within the framework of what Strawson calls the positivist view.

Strawson's description of phenomenal analyticity does distinguish his from the positivist view. He says, in ostensibly familiar terms, that phenomenal propositions are true in virtue of meanings. But these, unfamiliarly, are 'essentially phenomenal, visual meanings . . . essentially picturable meanings'. Proof in phenomenal geometry involves 'a phenomenal exhibition of meanings', in which 'phenomenal figure-patterns can be elaborated to exhibit an extensive system of relations between phenomenal spatial concepts'. [p. 286.]

We cannot, either in imagination or on paper, give ourselves a picture. . . . Such an impossibility used to be expressed by saying that such axioms are necessarily true because self-evident. This left the character of the necessity, or the impossibility, insufficiently explained. We can explain it by saying that the axioms are true solely in virtue of the meanings attached to the expressions they contain, but these meanings are essentially phenomenal, visual meanings, are essentially picturable meanings. Any picture we are prepared to give ourselves of the meaning of 'two straight lines' is different from any picture we are prepared to give ourselves of the meaning of 'two distinct lines both of which are drawn through the same two points'. [p. 283.]

This is not Nagel's positivistic unfalsifiability, and a positivistic account of truth in virtue of meaning (rules of use, and so forth) would render irrelevant the visual images Strawson emphasizes.

But where it is distinct, Strawson's account is elusive. The expressions 'phenomenal exhibition of meanings', and so forth, by themselves convey little beside echoes of the notion that an object such as a mental image could be the meaning of a term. Strawson's use takes us no further. He passes from, for example, 'picture showing two straight lines' to 'picture of the meaning of "two straight lines" ', without separating the two. This leaves us unable to distinguish a phenomenal exhibition of meanings from a plain exhibition of phenomenal figures.

This in turn leaves obscure the purported grounding of phenomenally analytic truths. For an exhibition of phenomenal figures, like one of physical objects, could naturally be taken to support no more than the claim that a certain geometry was contingently true of the exhibited objects. Again, the visualizing which Strawson calls a phenomenal exhibition of meaning Nagel calls an experiment in the imagination. It is hard to see why Strawson's descriptions should be preferred.[15]

The difficulty with Strawson's account at this point closely resembles that of Kant's account of the intuitional foundation of a synthetic *a priori* proposition. This is an exegetic felicity. Strawson intends his account to mirror Kant's. He hopes partly to vindicate Kant's belief that 'the construction of concepts in pure (i.e. non-empirical) intuition' is a source of geometric knowledge, by showing how 'Kant's theory of pure intuition can be construed as a reasonable account of the nature of geometry in its phenomenal interpretation'. (pp. 277; 283–4.) It is to this end that he compares proof by consideration of a phenomenal figure to Kant's construction in intuition. Now, the synthesis or empirical element in construction is the source of Kant's mathematical *synthetic*; so no wonder we feel it to conflict with Strawson's geometric *a priori*.

A belief that geometric proof can in part essentially be accomplished by construction or exhibition of a figure may explain some features of Strawson's and Kant's descriptions. If construction were part of demonstration it would perhaps be appropriate to speak of the construction or exhibition of concepts, meanings, or conceptual connections. If construction were essential to demonstration, if it could not be eliminated or replaced by statement, it would be difficult to distinguish proof from experiment, or exhibition of meanings from exhibition of objects.

Of course it seems that construction or exhibition can have no role at all in proof. By a proof we understand a set of statements, premises and conclusion. If the premises do not entail it, the conclusion is not yet proved, construction or no; if they do, nothing further is needed. So a construction or exhibition is either impotent or otiose.

More generally, it seems, as on the positivist view, that there is no reason to differentiate kinds of necessity among *a priori* propositions; and that we could find no foundation outside language for the necessity there is in proof. Consequently it seems there is no specific phenomenal necessity, nor could there be any such explanation of it as Strawson attempts to provide.

This means we find no explanation of phenomenal necessity either within or without the positivist view. There is no problem here. For there is no reason to accept the assumption, underlying both Strawson's account and that of Nagel and Reichenbach, that phenomenal propositions are necessarily true.

Rather it seems that we should take phenomenal geometry simply on a par with physical geometry, and hold that its propositions, if true, are contingently true. For, schematically, if putting 'physical straight line' for 'straight line' in a geometry produces a contingent theory about physical straight lines, then putting 'phenomenal straight line' should produce a contingent theory about phenomenal straight lines. Nothing in the nature of the case forestalls this.

We cannot picture two straight lines between two points. A colour-blind person, or one lacking certain experiences, may be unable to picture anything red, and no one can picture other than certain colours. We should presumably say these latter were empirical propositions, contingent on persons' experience or powers of discrimination. Why not the former?

Statements about the imagination may suggest the *a priori*. They are verifiable by introspection, not examination of the world. (Compare Strawson, p. 282.) No alternative to their truth can be imagined (in the sense that we cannot, even by trying very hard, imagine what in fact we cannot imagine; nor what another whose power exceeds ours imagines; and so forth). These resemble Kant's grounds for calling propositions *a priori* and give the designation a certain fitness. But as features of contingent propositions about imagination they cannot make a proposition *a priori* in any sense contrasting with the contingent or empirical.

Thus we can regard phenomenal propositions as contingent, and hence as fitting in the positivist framework. A further description of their contingency follows upon the resolution of the paradox with which we began.

VI

This paradox was that it seems we must picture space as Euclidean, whereas on scientific grounds we may judge it non-Euclidean. It seems odd that we cannot picture things as they are, that rather we are constrained to picture the contradictory of what we might have scientific reason to think true. This was exemplified by Strawson's acceptance of a necessarily Euclidean geometry of the visual, together with a non-Euclidean geometry of space. So it will be appropriate to begin by noting what he says about the latter.

The testing of Euclidean geometry by observation and measurement shows its theorems to be verified with an acceptable degree of accuracy for extents of physical space less than those with which astro-physics is concerned; but for astro-physics itself, a different physical geometry, incompatible with the Euclidean, is found to accommodate observations and measurements more simply. [pp. 285–6.]

The situation alluded to here is not that one geometry is true of small regions while another, inconsistent with it, is true of large regions. This could not be the case. Large regions are composed of small regions; and if one spatial region is Euclidean, and another adjoining region is Euclidean, then the larger region composed of the combined adjoining regions must also be Euclidean. So if we regard large regions as non-Euclidean, we cannot regard the small regions composing them as Euclidean. We must regard them as strictly, if undetectably, non-Euclidean.[16]

The fact is that the inconsistencies between the two geometries typically yield empirically detectable differences only in application to very large regions. In a small region considered in isolation, observation and measurement may fit equally well with either geometry. Here one can loosely say of either, as Strawson says of the Euclidean, that it is 'verified with an acceptable degree of accuracy'. At this degree of accuracy, however, one geometry is not verifiable in opposition to another which contradicts it. The purported verification is equally the verification of contradictory theories. Nevertheless, there may be good reason for regarding such a region as genuinely, if (locally) undetectably, non-Euclidean, as opposed to Euclidean. For it may be part of a large region which is detectably non-Euclidean. And this, as Strawson says, appears to be the case.

With this in mind, let us reconsider the assertion that phenomenal geometry is Euclidean. The relevant phenomenal figures are visual and mental images and the looks of things. The latter are Strawson's paradigms; their geometry is easy to determine. Thus phenomenal straight lines are 'the looks themselves which physical things have when, and in so far as, they look straight'. By 'physical things' here are meant 'physical lines or edges' examples of which are taut strings, light paths, and lines on paper. Three intersecting physical lines form a physical triangle, the look of which is a phenomenal triangle. The phenomenal triangle is Euclidean if the look of the physical one is; and this, presumably, is true if the physical triangle looks Euclidean.

It follows at once that phenomenal geometry is *not* Euclidean. For such a phenomenal figure as the look of a physical triangle is not. No physical triangle looks Euclidean as opposed to non-Euclidean. The difference made by the assumption that a physical triangle is Euclidean as opposed to non-Euclidean is visually undetectable. It therefore looks just as much non-Euclidean as Euclidean. Local observation and measurement fit equally with Euclidean and non-Euclidean assumptions; so it is not surprising that the looks of things fit equally with both assumptions. To say this is to say that these phenomenal figures fit both equally. Similarly for images: just as visually indiscriminable items have the same look, so the same image represents them indifferently. Phenomenal figures are therefore no more Euclidean than non-Euclidean. So phenomenal geometry is not Euclidean. Rather it is neutral or indeterminate.

(This line of thought requires the geometry of phenomenal items to be tied, as in Strawson's account, to the geometry that things are seen or imagined to have. Otherwise it is quite opaque what geometrical ascriptions to phenomenal items would mean, or how they could non-arbitrarily be made – let alone made with precision sufficient to differentiate geometries visibly indistinguishable in application. So it could hardly be argued that if the geometrical properties of phenomenal items were made independent, phenomenal geometry might still prove Euclidean.)

How, then, can we account for the plausibility of the suggestion that phenomenal geometry is Euclidean, and for the thought-experiments which seemed to establish phenomenal axioms?

Partly the explanation is simple. Euclid's geometry is familiar and approximately true. We naturally describe in familiar terms, and where measurement is concerned we correctly speak more or less imprecisely. We therefore naturally and correctly describe figures in Euclidean terms – just as, say, we call a line segment an inch long, despite the fact that its length may not be very precisely determinable, and not excluding the possibility that N such segments, where N is large, should produce a line greater than N inches long. 'Euclidean' here really means no more than 'approximately Euclidean'; and although 'Euclidean' and 'non-Euclidean' are contradictories, 'approximately Euclidean' and 'non-Euclidean' are here true together. It is easy to forget that approximation is involved and so to suppose, erroneously, that in this use 'Euclidean' contradicts 'non-Euclidean'. Hence a belief that things as they are look Euclidean, and that to look non-Euclidean they would have to be or look different, so as to fit contradictory descriptions. Or that our images represent Euclidean figures only, so that a different geometry would require different images.

Thus, for example, someone might try to picture a non-Euclidean triangle by starting with an image of a triangle assumed to be of 180 degrees, and trying to increase an angle without bending a side. This would prove somewhat frustrating; and it would be to overlook the fact that an image determines no exact angular sum for an imagined triangle. Since points and lines can be pictured only in terms of areas or their (imperfectly determinable) boundaries, any geometric image will be ambiguous. Here the same image can equally represent invisibly dissimilar alternatives, Euclidean and non-Euclidean. So no further picturing is required.

The indeterminacy may be overlooked also because of unreflecting exaggeration of the ability to picture. We can picture what we cannot see, either the very small or the very large, far or near. The capacity may seem unrestricted by size or distance; we can choose whatever scale for our pictures we please. One may feel, for example, that it should be possible to draw or imagine any simple figure, of any size, in space; and hence to picture figures on the astral scale, or with regard to the minute differences, relevant to the verification of non-Euclidean geometry.

One may also feel that we ought somehow to be able to form pictures, different from any we have, of the non-Euclidean; or, failing this, that our imagery is unambiguously Euclidean. If it is possible to picture with a precision or on a scale relevant to detecting non-Euclidean phenomena, it should be possible to picture detectably non-Euclidean phenomena; and our imagery remains as Euclidean when constructed with reference to an astral scale as when referred to the middle-sized or the very small.

In fact we simply do not picture relevantly here. It is true we can imagine or draw what we cannot see; but what we can imagine accurately, or picture accurately in general, has limitations connected with sight.

Suppose, for example, we wish to represent two stars and the distance between them, by dots and a blank space on a sheet of paper.[17] The dots can be related in circumference as the stars. But if the stars are sufficiently far apart in relation to their size, we will be able to form no picture in which their size is shown accurately in relation to their separation – in which, that is, the size of the dots is to the distance between them as the size of the stars is to the distance between them. For the dots may be so related that from any given point, if they are large enough to be seen then they will be so far apart that both cannot be seen at once. So if dots and distances are in scale, the picture cannot be taken in. If we want a picture which like a mental or visual image can be taken in at once, we can make it only by enlarging the dots in relation to their separation. The picture will then show stars larger in relation to their separation than those we set out to represent. Here, owing to the imperfection of sight, the only picture we can have is out of scale.

Or consider a very long pair of straight railroad tracks, to be pictured, as from above, by parallel lines. The rails will be a few inches wide and a few feet apart, but thousands of miles long. No picture will show us their width and separation in relation to their length. In no picture, that is, will the relation of length, thickness, and separation of the lines be the same as those of the rails. We are not capable of seeing lines related in length and thickness as such rails; we can see only relatively thicker lines. Consequently, any picture we can take in will show lines thicker in relation to their length than the rails we set out to picture; and similarly the separation of the rails will be shown out of proportion to their width or length.

A simple principle is involved. If a picture is to be taken in, the elements (for example, dots, lines) which compose it must be simultaneously visible. They will therefore have certain spatial properties and relations. Scale pictures like geometric diagrams show spatial situations by the spatial characteristics of their elements. Those characteristics required by considerations of scale may conflict with those needed for visibility. A distortion results from the sacrifice of scale to visibility. Similarly for images. Just as, say, there will be a maximum ratio of length to thickness consistent with the visibility of (the representation of) a line, there will be such a relation for any visualized line.[18] And as the maintenance of this ratio for visible pictures means that certain spatial relations cannot be pictured accurately, for images it means that they cannot be imagined accurately. This systematic possibility of distortion entails, among other things, that the ambiguity of images between Euclidean and non-Euclidean cannot be resolved by change of scale.

Someone may, for example, think he can picture Euclidean parallel straight lines. For simplicity, and to fix what is meant by a line, suppose he pictures such a pair of lines as could be drawn on a blackboard, a few feet apart and a few yards long, at the maximum ratio of length to thickness. Now it can be pointed out that his picture of these lines does not differ from one of lines which would meet if extended, say for a few miles. The picture does not exclude this possibility, so it does not show the lines as parallel. He may reply that he can regard the lines as extended; he can exclude the possibility that the lines he pictures would meet if extended, by picturing them as long as he likes. This is really the assertion that he can change the scale of his image to represent longer lines. But as the scale is changed, the picture ceases to show the disposition of *lines*. As the length represented increases so does the width and hence the area shown covered by what was to be a line; and nothing in the changed picture will be capable of showing how lines such as could be drawn on a blackboard are disposed. If a picture is to show lines of a certain kind its scale must be limited; if its scale is limited it cannot show the lines as parallels, or in general as Euclidean. So pictures, like sight, remain geometrically indeterminate, whatever our intentions as to their precision and scale.

In fact the limits of accuracy in imagination seem directly tied to those of picturing by visible pictures, and so indirectly to sight. Roughly, we should not expect to find a person capable of imagining a spatial situation accurately unless he was, or had been, capable of seeing an accurate visible representation of it. If a person were unable to see any pair of dots related in size and separation as a pair of stars or bits of dust he was attempting to visualize, we should expect to find that he could not visualize their size and separation accurately either. If he said he could not, the matter would presumably be settled. It would puzzle us if, knowing what was involved, he said he could; and in default of very special testimony, we should have no reason to accept his claim – we should reject it, or not know what to do with it. From this it seems we are justified in assuming that what persons can imagine accurately is limited to what they can see accurate pictures of, which in turn is determined by their powers of sight.

VII

The limits of geometric imaginability and their connection with contingency and necessity can now be more fully set out. The main points are perhaps as follows.

It seems (again roughly) that if a person can see items as of a certain kind, there is reason to accept his claim so to picture them. So if things are seen as they are, phenomenal geometry will depend upon how things are and how precisely they can be seen. The phenomenal geometry of someone able simply to see the non-Euclidean character of our space would be accordingly non-Euclidean. That of someone with perfect sight in a Euclidean world would be Euclidean; and of someone even with imperfect sight but in a visibly non-Euclidean world, non-Euclidean. As stressed, the geometry of imperfect sight in an unobvious world will be indeterminate.

These other phenomenal geometries are, explicably, not ours. As has been shown we cannot form the non-Euclidean pictures of our space we might have with more powerful sight. Nor can we simply alter our way of seeing distances and shapes to what it might be in a different world. But now this latter inability can be described a bit further by reference to some abstract features of geometry.

Determining the geometry of space requires comparing distinct spatial intervals. This is typically thought of as accomplished by the use of standards of length, such as a portable rod which realizes a certain interval and whose coincidence relations with other objects and intervals provides their measure. Measurement then consists in the establishing of these coincidence relations.

A given set of coincidence relations among rods, objects, and intervals generally can be interpreted in terms of measure and geometry in various ways. In particular, the relations can yield one set of measurements and one geometry if the interval realized by a standard rod is taken as everywhere the same, other measurements and another geometry if the interval is taken to vary with the position and orientation of the rod. The differences in measurements will result in the relations' determining different sets of intervals congruent or equal. And with one set of intervals congruent the geometry will be Euclidean, with another, non-Euclidean.[19]

Since this indeterminacy arises in interpreting the facts of coincidence on the basis of which measurements are assigned and geometry assessed, it cannot be resolved by any further recourse to measurement or geometry. Still, the coincidence relations themselves may fix the geometry, by practically ruling out the assumption that the length of the standard varies. For to retain the same size can be little more than to retain the same size in relation to things in general. So if coincidence relations among the standard, bodies, and items with size in general are unvarying, the standard is (to be regarded as) rigid. Given such rigidity, congruence becomes simply coincidence with a standard.

(Here, as one might say, the harmony of things with size can fix geometry despite indeterminacy. Einstein was inclined to assimilate the non-Euclidean geometry of a gravitational field to such a case,[20] and philosophers of science, among others, have followed him. But the cases are not entirely comparable. For in Einstein's theory a gravitational field changes the coincidence relations – the relative shapes and sizes – of bodies of different shape and size.[21] In the field things are non-Euclidean measured by small rods where these are taken as rigid. But no harmony of coincidence relations forces us to take small rods as rigid. Consequently the choice between geometries must be made on less obvious grounds.

Perhaps Einstein can be seen partly as giving the simplest account of local measurement and its most direct embodiment in geometry. Given the fundamental role of local measurement in Einstein's physics and in the verification of physical theory generally, this procedure seems appealing; its justification would be complex.)

Now as an expression of the indeterminacy of geometry, we have

(a) It is possible to describe a world as Euclidean or non-Euclidean, depending upon which of its intervals are taken as congruent or equal

while also a world will be fixed as Euclidean or as non-Euclidean if found so by measurement using standards whose relative size, like the relative size of things in general, stays constant. And this is not arbitrary: the bodies, and so forth, of such a world are to be regarded as rigid on conceptual grounds.

Still, in consequence of *(a)*,

(b) It is possible to describe a non-Euclidean (Euclidean) world of rigid
bodies as a Euclidean (non-Euclidean) world of bodies changing
dimensions with position and orientation, but in such a way that their
coincidence relations stay constant.

Since they stress only the relations of geometric descriptions, these principles
might loosely be called logical. Now suppose we apply them to the descriptions under which things are seen, and so to visual geometry, by putting 'see'
and 'seen' for 'describe' and 'taken' in *(a)* and 'see' for 'describe' in *(b)*. We
then have modified, visual principles, to the effect *(a)* that it is possible to
see things as Euclidean or non-Euclidean depending upon which intervals
are seen as congruent and *(b)* that it is possible to see a non-Euclidean rigid
world as Euclidean changing and vice versa.

These visual principles could be variously interpreted. The possibility of
seeing something under a description might be taken simply as given by its
being so describable. The visual principles would then be versions or
rephrasings of *(a)* and *(b)*, and hence (loosely) logical statements. Or they
might be taken as substantive claims – for example, about persons' abilities
to see things as falling under alternative geometric descriptions. Here the
principles would be contingent statements, and possibly false.

Now Reichenbach's discussion of geometric visualization pivots on such
principles. He says, in accord with *(a)*:

Space as such is neither Euclidean nor non-Euclidean . . . it becomes Euclidean if a
certain definition of congruence is assumed for it . . . if a different definition of
congruence is introduced . . . space becomes non-Euclidean.[pp. 56–7.]

He holds that we visualize with a Euclidean definition of congruence.
This is only because our world is so nearly Euclidean: if things were different
our way of seeing would change in accord with the possibilities of *(b)*,
interpreted visually:

if in daily life we dealt occasionally with rigid bodies that adjusted themselves to
non-Euclidean geometry . . . At first we would have the feeling that objects *changed*
when transported . . . After some time we would lose this feeling and no longer
perceive any change . . . we would have adjusted our visualization [to a non-
Euclidean geometry]. [pp. 54–5.]

This is an example of the essential change. Since we are adjusted to
Euclidean congruence, and since non-Euclidean visualization means visualization adjusted to a non-Euclidean congruence, we need only undergo
such a change of sight or visualization, such an adjustment of congruence, to

accomplish non-Euclidean visualization. Thus Reichenbach gives a principle of visual adjustment of congruence corresponding to *(a)*:

Whoever has successfully adjusted himself to a different congruence is able to visualize non-Euclidean structures as easily as Euclidean. [p. 55.]

He takes this as supporting a substantive claim:

The mathematician is thus correct in saying that he has become accustomed to visualize non-Euclidean geometry. [p. 58.]

This he applies, as we saw earlier, to the interpretation of geometric drawings.

Although this account of what is involved in non-Euclidean visualization requires a number of assumptions (for example, about the identification of images, the propagation of light, and so forth) it is appealing and seems informative. But it does not support Reichenbach's belief that persons can actually visualize non-Euclidean geometry. Neither the fact that our way of seeing depends upon how things are, nor reasonable speculation about how we should adapt if things were different, shows that *as things are* we can see or visualize in any other than the familiar approximately Euclidean (or weakly non-Euclidean) mode. We cannot.

For, as I have argued, the change in visual congruence on which this account of non-Euclidean visualization pivots does not occur. No one in fact experiences a change of sight relevant to seeing or visualizing in non-Euclidean terms. It seems in consequence that those who claim non-Euclidean visualization do not actually accomplish it. Rather they visualize in familiar terms while describing their images non-Euclideanly.

(This is easily recognized in one of Reichenbach's own examples. He says the small drawing known as Klein's model of a non-Euclidean space can be

'truly a visualization of Lobatchewsky's space' since 'it is possible to adjust to the other congruence'. But

in order to accomplish the visualization, we must 'forget' everything outside the circle . . . We must imagine ourselves inside the circle and remember that the periphery cannot be reached in a finite number of steps. [p. 58.]

There is no telling in what such a visualization might consist – to what visual image could such descriptions (involving infinity, being inside, and so

forth) meaningfully be applied? and how might it relate to the drawing? –
and we simply have no idea of a change of sight which might accomplish it.
Really, Reichenbach must picture the circle just as we do. Hence the notion
of change to an alternative visualization, and with it the notion of non-
Euclidean visualization, has been given no content. Its emptiness is perhaps
hidden by the ornamentation of Reichenbach's analysis.)

This conclusion might of course be refuted by the testimony of visualiz-
ers; but so far as I know, no testimony of any weight has been given.

VIII

There remains the fact which caught the attention of Bennett and
Strawson – that we cannot picture two straight lines between two points.

It may seem obvious that since the visibly determinable spatial features of
objects fit both geometries, appearances must also; and it follows naturally
that picturing will be consistent with both.

Some images are clearly neutral in this way. It is easy to regard an image
or picture of a triangle, for example, as consistent with both geometries and
hence as showing either equally. But no image shows equally the Euclidean
situation of one straight line and the non-Euclidean situation of two straight
lines, between two points. The pictures we have on the face of it show only
the Euclidean phenomenon.

This perhaps explains the peculiar impression of intuitive self-evidence
associated with the axiom; and its consequent central role in producing the
conviction that phenomenal geometry is Euclidean.[22] But it is itself explicable in terms of the kind of distortion in picturing encountered already. In
consequence it can be seen to have no bearing, either on the axiom or on the
claim that phenomenal geometry is Euclidean.

Consider the situation Barker describes: between two very distant points
are two paths measuring equally short and shorter than any other, which
light rays follow and along which taut cords lie.[23]

Suppose we wished to picture it. One way, illustrating the principle
involved, would be to stretch a suitably large sheet of paper between the
points and make a picture by drawing along the lines. Now clearly this
picture could not be seen. Someone far enough away to see both points
would be too far off to see the lines, which would be minute in comparison
to the sheet. The only way to make anything visible here would be to thicken
the lines. But then they would overlap before becoming large enough to be
seen. So *two* lines could not be seen. Owing to the distortion required to
make the lines visible, the only way to make two lines visible would be to

bend one away from the other. Then one line would be and appear curved. Hence the only usable (visible) pictures fail to show two lines, or show one curved. The same is true, for like reasons, of our images and other pictures.

So really there is no accurate picture of the situation described. Paths of the required ratio cannot be pictured. Because of their relative thickness, the areas which can be pictured cannot mirror the disposition of lines; and in this case the particular form of distortion leaves no alternative but pictures easily interpreted as showing Euclidean lines. It is like the transformation of a delicate design painted over with a thick brush.

Since no picture here is capable of showing the disposition of lines in space, none shows these lines as Euclidean. Just as no picture could show two straight lines between two points if they were there, so no picture shows the one and only straight line there is. At this scale and in this case we can only disregard our images; we cannot take them as showing how things are. So despite the impression, our images are not really Euclidean; rather they are too crude to serve.

IX

So, finally, nothing constrains us to picture space in terms of a superseded theory. The impression is only the result of misleading pictures. We can neither picture every spatial situation nor change our way of picturing at will; but still we see and picture consistently with Euclidean and non-Euclidean theories. Possibly this is not obvious, but it ought not be surprising. It has always been clear that the observations required to tell between physical geometries could not be made by unaided sight.

[1] Frege, *The Foundations of Arithmetic*, trans. by Austin (Oxford: Blackwell, 1950), 20.

[2] Jonathan Bennett, *Kant's Analytic* (Cambridge University Press, 1966) 31. Bennett's position is more complex than might be inferred from the passage quoted. He is opposing 'that preoccupation with the visual which has weakened and narrowed epistemology for centuries', and he says of the passage, 'I am not sure this is right, perhaps because I am not sure what I mean by "must look curved" '.

[3] P. F. Strawson, *The Bounds of Sense* (London: Methuen, 1966), 283. All quotations from Strawson are from this book, and only their page number is cited in the text. This view illustrated by Strawson and Bennett has been surprisingly common. See Ewing's *A Short Commentary to Kant's Critique of Pure Reason* (London: Methuen, 1938), p. 45, for his statement and that from Johnson's *Logic*; also the text and notes to 'Empiricism and the Geometry of Visual Space' in Grünbaum, *Philosophical Problems of Space and Time* (New York: Knopf, 1963), for references to Carnap and others. W. and M. Kneale seem to discuss it in *The Development of Logic* (Oxford: Clarendon Press, 1962), pp. 385f.

[4] Kant, *Critique of Pure Reason*, trans. by Kemp Smith (London: Macmillan, 1929), A713/B741. Compare Mill, *System of Logic* (London: Longman, 1843), II, v. 5: 'The foundations of geometry would therefore be laid in direct experience, even if the experiments (which in this case

consist merely in attentive contemplation) were practised solely upon what we call our ideas, that is, upon the diagrams in our minds.'

[5] Frege, *Foundations*, 101f. According to Reichenbach the shadow of this view remains.

'The relativity of geometry has been used by Neo-Kantians as a back door through which the apriorism of Euclidean geometry was introduced into Einstein's theory: if it is always possible to select a Euclidean geometry for the description of the universe, then the Kantian insists that it be this description which should be used, because Euclidean geometry, for a Kantian, is the only one that can be visualized.' (*A Einstein, Philosopher-Scientist*, ed. by Schilpp [Evanston: Library of Living Philosophers, 1949] I 299.)

On the relativity of geometry see below, pp. 58–9, the argument of the paper shows the irrelevance of this Kantian insistence.

[6] In Einstein *et al.*, *The Principle of Relativity* (London: Methuen, 1923), 161.

[7] S. F. Barker, in the *Encyclopedia of Philosophy* (New York: Macmillan Co., 1968), III, 288. I *think* there are not multiple paths between every pair of points, but only certain pairs. Also, Barker is describing only one kind of field. In others the metrical situation is neither so definite nor so direct an extension of everyday technique. According to Reichenbach there are gravitational fields in which the geometry given by light rays differs from that given by rods; and in some fields in which there is not a unique light path between points, the metrical situation is so indeterminate it hardly seems useful to speak of lines in an ordinary sense at all. (See *The Philosophy of Space and Time* [New York: Dover, 1957], ch. 27; all textual references to Reichenbach are to this book, and only their page number is given.)

[8] See Reichenbach, esp. chs. 9, 10, 11, 13. He does not discuss the problem mentioned above, although sometimes he seems close (pp. 47, 91) and provides material for its solution.

[9] Frege, *Foundations*, 20.

[10] J. R. Lucas, *British Journal for the Philosophy of Science*, **20** (1969), 6. On one interpretation of what Lucas says, the case is comparable to that treated by Strawson at pp. 290–1. Lucas also distinguishes the space of our ordinary experience, which presumably we inhabit, from that which our physical theories are about. But he gives no reason for this distinction, and on the face of it, it is implausible.

[11] Craig, *British Journal for the Philosophy of Science*, **20** (1969), 121–34, explains the use of projection in this context. 'Suppose we are looking at some figure. Whatever it may be, and irrespective of whether it "looks Euclidean" or not, there will be a projection of it, as we see it, on to a Euclidean plane. . . . So, whatever the nature of the figure, our sense impressions of it could, by this criterion, be said to be Euclidean, or, at any rate, not to be non-Euclidean. This is a necessary proposition; but a very weak one.' [pp. 132–3]. I do not wish to claim that there could be no use for 'looks non-Euclidean' or that no picture whatever might some way be describable as showing two straight lines intersecting twice. Escher's drawings, as Craig points out, might be said to look geometrically or logically impossible, or to show impossible situations. I don't know what pictures like this a clever artist might produce, or what arguments a philosopher might give to support that description. But I think such a case will be distinguishable from those I discuss.

[12] Nagel, *The Structure of Science* (London: Routledge and Kegan Paul, 1961), 225. Nagel gives a lucid account of what Strawson calls the 'positivist view' of geometry. He and Reichenbach may have felt that only in this way could the Euclideanness of phenomenal geometry fit the positivist view.

[13] Ibid.

[14] Someone might wish to urge that the proper description of some seen or imagined lines was that the two looked (were imagined) equally short and shorter than any alternative. Here one could hardly use the Euclidean axiom to force the judgement that one or both must look curved, or say that the image had been constructed in accord with the axiom; for both are conceded to look or be straight (short) in Euclidean terms, and equal. Such a description fits with the indeterminacy thesis argued below, p. 54ff. It might be claimed that someone who refused to give this non-Euclidean description in an appropriate case was using the Euclidean axioms implicitly; but this would be difficult to establish.

[15] Another move to meaning in this case is quoted by Mill from Bain (*System*, II, v. 5): 'We cannot have the full meaning of Straightness, without going through a comparison of straight objects among themselves and with their opposites, bent or crooked objects. The result of this

comparison is, *inter alia*, that Straightness in two lines is seen to be incompatible with enclosing a space; the enclosure of space involves crookedness in at least one of the lines.' The idea that understanding meaning involves being able to treat cases is a good one; but the feature here attributed to meaning was surely induced from the cases.

[16] Lucas argues that 'it is a necessary condition of our being able to apply the concept "same shape though different size" that our geometry should be Euclidean', and that hence 'the price of abandoning Euclidean geometry would be the loss of an important respect in which things can be similar to or dissimilar from one another . . . we should no longer be able to classify by shape.' But clearly we can regard Euclidean geometry as false and still take things as comparable in respect of shape and indeed, for all practical purposes, as having the same shape but different sizes. It suffices to regard things as (non-Euclidean and) approximately Euclidean. The strength of the approximation, in fact, makes the Euclidean concepts as usable as any.

The facts and connections Lucas cites do not prove his contention that we must regard things as Euclidean. Together with the approximate truth of Euclidean geometry, however, they partly explain its outstanding naturalness and historical pre-eminence. Some such explanation is surely better than Strawson's in terms of phenomenal necessity.

[17] I explicitly treat only two-dimensional pictures seen from straight on. This seems adequate to account for visualization of geometric figures, Mill's 'diagrams in the mind.' I think analogous considerations would apply to other kinds of pictures and models.

[18] Lines of this kind are components of the most familiar geometrical diagrams. You might have another kind of visual geometry, say using areas shown by colour patches, and represent lines by colour edges. Similar considerations regarding accuracy still apply. There will be limits on the kind of colour areas visualizable and the indeterminacy of the visible location of a colour edge will mean that it can be treated as I treat (areas representing) lines here.

[19] See Grünbaum for exegesis of these matters.

[20] *The Theory of Relativity* (London: Methuen, 1920), 85–6. Einstein's simplified analogue treats only of the coincidence of rods and so ignores other bodies.

[21] See Swinburne, *Space and Time* (London: Macmillan, 1968), 92–3. I do not think his description of 'the original interpretation' of the general theory applies to the paper of Einstein's to which he refers.

[22] This axiom is usually given as a likely candidate for intuitive self-evidence. See, e.g., Einstein, in Feigl and Brodbeck, *Readings in the Philosophy of Science* (New York: Appleton-Century-Crofts, 1953), 189.

[23] Only very long lines would be empirically distinguishable. But presumably the argument would apply locally as well.

III

THE PROOF-STRUCTURE OF KANT'S TRANSCENDENTAL DEDUCTION

DIETER HENRICH

THE transcendental deduction of the categories is the very heart of the *Critique of Pure Reason*. It contains the two principal proofs of the book, the one demonstrating the possibility of a systematic knowledge of experience and the other the impossibility of knowledge beyond the limits of experience. Kant himself considered this theory completely new and extremely complicated; moreover he conceded that he had great difficulty in working out a satisfactory exposition of his theory. It is one of the two chapters which he rewrote completely for the second edition. Thus it is not surprising that this deduction has preoccupied interpreters more than any other text in the history of philosophy. In only thirty-five pages, which are easily separated from their context, Kant has formulated his most profound thoughts and presented the decisive foundation for his theory of knowledge. Whoever understands these pages possesses a key to the understanding and evaluation of the entire work. But Kant's text is so complex and elusive that it is difficult to follow the line of argument and to perceive within it the structure of a proof which could support the whole system of critical philosophy. In view of this it has been easy for Kant's critics to focus their attacks on the deduction. By the same token it has been just as easy for philosophers who wish to make use of Kant as testimony to their own position, to read their thoughts into his. Until now, however, no one has been able to offer an interpretation which agrees fully with Kant's text.

Hence, there is still controversy over which of the two versions of the deduction deserves priority and whether indeed any distinction between them can be maintained that would go beyond questions of presentation and involve the structure of the proof itself. Schopenhauer and Heidegger held that the first edition alone fully expresses Kant's unique philosophy, while Kant himself, as well as many other Kantians, have only seen a difference in the method of presentation.

From *Review of Metaphysics* 22 (1968–9), pp. 640–59. Reprinted with permission of the *Review of Metaphysics*.

In the following, an attempt will be made to settle this conflict which has persisted more than 150 years.[1] We shall advocate the thesis that only the second edition develops a tenable argument and that the argument in this version corresponds more adequately with the specific structure of Kant's thought as a whole, than does that of the first edition. This position contradicts the most important interpretations of Kant; moreover it proposes to re-evaluate the meaning of his work and to guide its reception in a direction other than that of speculative Idealism, Neo-Kantianism, or Existential Philosophy.

I

We will treat first another controversy which, compared with the debate over the value of the two editions, is only of minor importance, yet which is relevant here insofar as it ultimately leads back to this question and allows it to be answered: namely, the controversy concerning the structure of the proof in the second edition.

In this edition the conclusion of the deduction seems to be drawn twice in two completely different passages. It is the task of a transcendental deduction to demonstrate that the categories of our understanding are qualified to provide knowledge of appearances, as they are given to us in the unity of a synthesis of experience (B 123). The conclusion of section 20 reads: 'Consequently, the manifold in a given intuition is necessarily subject to the categories.' (B 143.) This conclusion does not seem to differ from the result of section 26, according to which 'the categories . . . are . . . valid a priori for all objects of experience' (B 161).

Thus one is tempted to see two proofs of the same proposition in the text of the second edition. That leads, however, into direct conflict with Kant's unequivocal explication in section 21, which states that two arguments, rather than two proofs, are involved and that these together constitute the proof of the deduction. 'Thus in the above proposition a beginning is made of a deduction of the pure concepts. . . .' 'In what follows, [something further] will be shown. . . .' 'Only thus, by demonstration of the *a priori* validity of the categories in respect of all objects of our senses, will the purpose of the deduction be fully attained.' (B 145.) We can now formulate a criterion for a successful interpretation of the whole text of the deduction in this way: the interpretation must show that, contrary to the initial impression that the two conclusions merely define the same proposition, on the contrary, sections 20 and 26 offer two arguments with significantly different results, and that these together yield a single proof of the transcendental deduction. We shall call this task the problem of the two-steps-in-one-proof.

In previous commentaries this problem has been either pronounced insoluble or else passed over in silence.[2] The better commentaries claim that Kant's assurance that his deduction presents two steps in one proof cannot be taken seriously, and that we are compelled to read the text as two distinct and complete proofs. Two proposals made on the basis of this double-proof theory merit our attention.[3] We shall call them the interpretation according to Adickes/Paton[4] and the interpretation according to Erdmann/de Vleeschauwer[5] and shall examine them in that order.

1. In the preface to the first edition of the *Critique*, Kant himself distinguished an objective and a subjective side of the deduction (A XVI). The objective side makes the validity of the categories intelligible, the subjective investigates their relation to the cognitive faculties in us which must be presupposed if these categories are to be used. According to Kant one can also distinguish these two aspects as the demonstration *that* the categories have validity, and the demonstration *how* they attain validity. Adickes and Paton propose that this distinction be employed in order to understand the division of the deduction into two arguments: section 20 completes the proof of objective validity, section 26 demonstrates the subjective conditions of application.

This proposal has the advantage of being able to invoke in its support certain fundamental Kantian statements about the deduction – but there is no further evidence for it. For it is clear that the proposal cannot be applied to the structure of the second version of the deduction. In section 21 Kant clearly stated that the demonstration of the validity of the categories would be completed in section 26 (B 145). The title and conclusion of this section can be read in no other way. And the text itself contains no reflections about the interconnection of our cognitive faculties. The little word 'how,' which can indicate the distinction between a psychological and an epistemological investigation, a subjective and an objective deduction, only appears incidentally. In this context, however, we shall see that it must be understood quite differently.

2. The proposal of Erdmann and de Vleeschauwer likewise attempts to understand the second version of the transcendental deduction with the help of another observation of Kant's – this time of a distinction made in the first version of the deduction. In two corresponding trains of thought, Kant here elaborates the relation between the categories, which can be developed from self-consciousness, and the given sensible representations. He distinguishes them as the demonstration 'from above' and that 'from below'. In this way he implies a hierarchy of cognitive faculties, the highest

of which is the understanding and the lowest sensibility – extremes between which the faculty of imagination establishes a relation of possible coordination, and between which the two proofs move in opposite directions.

It seems quite natural to apply this distinction to the interpretation of the second edition. And indeed Erdmann and de Vleeschauwer propose that section 20 be understood as a deduction 'from above', while section 26 is to be regarded as a deduction 'from below'.

This proposal is in better agreement with the text of section 26, which has supplied the decisive arguments against the interpretation of Adickes and Paton. For Kant here actually proceeds from intuition, mentions the achievement of the faculty of imagination, and comes then to speak of the unity in the forms of intuition, which can be reached only through the categories and by virtue of the unification of the manifold in a consciousness (B 160). Nevertheless the two parts of the deduction remain unexplained by this proposal for the following reason: the structure of the first argument in section 20 can in no way be conceived as a deduction 'from above' – and thus as a process which differs from the argument of section 26 in so far as its proof must be constructed in the opposite sequence. In section 20, just as in section 26, the manifold of a sensible intuition is mentioned first. Then it is shown that the manifold can assume the character of a unitary representation only if it is subject to the categories. Thus both arguments establish that a given intuition can become a unitary representation only when the intellectual functions of the understanding are applied to it. Now as to whether or not this argument can properly be understood as a deduction 'from below': the forms of these proofs in no way make it possible to draw a meaningful distinction between the considerations of the two sections.

Hence the failure of the only proposed interpretations – not only because they depart from Kant's assurance that there is *one* proof presented in *two* steps and attempt instead to find two distinct proofs, but also and primarily because their arguments can offer no useful explanation of the distinction between the two proofs.

We must search for another interpretation of the text. It should avoid both of these errors as far as possible and seek an understanding of the proof of the deduction that would require the two-steps-in-one-proof thesis. Moreover, it cannot derive support, as do the proposals just discussed, from Kant's observations about the structure of the proof of the deduction, for they are valid only in the context of the first edition. Kant always allowed so many different trains of thought to influence him in formulating his central arguments that it is never possible to employ his comments unless he has explicitly related them to the passage of the text in question.

II

But now from the propositions of sections 20, 21, and 26, we can develop a proposal which will solve the problem of the two-steps-in-one-proof. Its plausibility stems from the fact that it makes intelligible many peculiarities of the text which must be neglected by all other proposals.

Kant obviously attached importance to the fact that the result of the proof in section 20 contains a restriction: he established that intuitions are subject to the categories *in so far* as they, as intuitions, already possess unity (B 143). He indicates this restriction very clearly by writing the indefinite article in the expression 'in an intuition' (*in Einer Anschauung*) with the first letter capitalized. Norman Kemp Smith, the translator, has misunderstood this hint.[6] He believes that Kant wanted to imply that some single intuition was subject to the categories. This interpretation, however, leads to no meaningful emphasis in the course of the proof. Unlike English, in German the indefinite article (*ein*) and the word unity (*Einheit*) have the same root. This made it possible for Kant to express through the capital letter not the distinctness of any arbitrary intuition as opposed to others (*singularity*), but rather its inner *unity*.

The result of the proof in section 20 is therefore valid only for those intuitions *which already contain unity*. That is: wherever there is unity, there is a relation which can be thought according to the categories. This statement, however, does not yet clarify for us the *range within which* unitary intuitions can be found.

The restriction of the proof in section 20 is then discussed in that part of section 21 which makes reference to section 26. Here it is announced that the restriction just made in section 20 will be overcome in the paragraphs of section 26, i.e., the second part of the deduction will show that the categories are valid for *all* objects of our senses (B 161). And this is what actually takes place. The deduction is carried out with the help of the following reasoning: wherever we find unity, this unity is itself made possible by the categories and determined in relation to them. In our representations of space and time, however, we have intuitions which contain unity and which at the same time include *everything* that can be present to our senses. For indeed the representations of space and time have their origin in the forms of our sensibility, outside of which no representations can be given to us. We can therefore be sure that every given manifold without exception is subject to the categories.

At this point the aim of the proof of the deduction has been attained, in so far as the deduction seeks to demonstrate the *unrestricted* validity of the categories for everything which can be meaningfully related to experience.

Perceptions, which arise erratically and which cannot be repeated according to determinate rules, would not make intelligible a coherent and systematic knowledge of experience. The only conceivable result of a limited capacity for ordering our sense-data would be a diffuse and discontinuous sequence of perceptions.

It is certainly extraordinary to claim that our capacity for making conscious and thereby unifying our own sensuous representations could perhaps only be limited. However, its conceivability is an immediate result of the fundamental argument of the whole *Critique*. It is implied that our consciousness has the peculiarity of being 'empty'. Everything of which we can become conscious must become accessible to us through media which do not immediately depend on this consciousness. According to Kant, it is for this reason that consciousness must be understood as an activity, thus always a *making*-conscious whose necessary inner unity causes us to give it the name 'I'. But this activity always presupposes that something is present in the first place which is to be made conscious. Thus our consciousness can be found only together with a 'passive', receptive faculty, which is distinct and in certain respects opposed to the spontaneity of consciousness; it can encounter intuitions only as given 'before all consciousness'. Kant reformulates the task of the transcendental deduction with reference to this very distinction: it must demonstrate that categories are capable of taking up something given into the unity of consciousness. 'Appearances might very well be so constituted that the understanding should not find them to be in accordance with the conditions of its unity.' (B 123.) If that is possible, then it can also be asked whether such a disproportion between consciousness and givenness can be excluded for all or only for part of the given appearances. The difference between these two possibilities also defines the difference between the result of the proof of the first and that of the second step of the deduction.[7]

III

This question need not occur at every level in the analysis of the conditions of our knowledge. It could be that considerations are possible such as would establish rather quickly that the alternatives with which the transcendental deduction has to deal are not three-termed but rather only two-termed: that therefore either *no* sensuous representations or else *all* sensuous representations are capable of being determined by the categories. Anyone familiar with Kant's work will suspect that Kant had good reason to assert this. But this amounts to saying that Kant also had an alternative way of constructing the proof of the transcendental deduction, other than the one which he actually used in the second edition. For in this construction he takes into account the possibility of a merely partial ability of the understanding to establish unity in the sensible representations. He excluded it only because we do in fact have unitary representations of space and time and therefore can also unify all representations of sense.

Fortunately we can demonstrate that Kant himself was actually conscious of the fact that the transcendental deduction could also be constructed quite differently. His pupil Johann Sigismund Beck undertook in the year 1793 to publish a selection from Kant's writings.[8] On the title page he was able to announce that this was being done on Kant's own advice. Kant was interested in making available a competent commentary which could also be used for lectures. But when Beck published the third part of his selections in the year 1796, he considered it necessary to undertake a fundamental investigation in order to specify the standpoint from which Kant's *Critique* was actually to be evaluated. He had come to the opinion that the structure of the book promoted a false estimate of Kant's doctrine. Thus it would be necessary to begin with the productive activity of the understanding, in order to avoid the misunderstanding that Kant really wanted to speak of 'given concepts' and of 'objects which affect us'. In Beck's opinion all this talk was only an accommodation to traditional doctrine and constituted preliminary concessions for the purpose of an introduction into the system. With the interpretation, Beck approached, somewhat belatedly, Fichte's philosophical conviction.

Naturally Kant could not bring himself to approve this. But since he was interested in Beck and in the effect of his writings, he was more willing to consider Beck's proposed alteration of the *Critique* than was his custom in comparable cases. In a letter to Beck's colleague Tieftrunk, he tried to show approximately what form the *Critique* might assume in an altered presentation.[9] Thus we see that Kant himself at one time proposed an alternative to the transcendental deduction of the second edition.

It must begin with the doctrine of the categories as rules for the unity of a possible universal consciousness – corresponding to sections 16–18 of the second edition. Then it must demonstrate that intuitions *a priori* are *presupposed* in order that the categories can be applied at all to given sensuous intuitions. This becomes evident, when one considers that the categories can only be conceived as operators that don't indicate in themselves the conditions under which they can be applied. Without such a possibility of application an essential moment of their meaning is missing. The meaning of *a priori* concepts such as the categories can only be accessible *a priori*. But the only possibility of securing a meaning *a priori* for the categories is their application to a form of sensible intuition – the only *a priori* element which is conceivable in the domain of their application to sensible givenness. If there is no *a priori* intuition, then there is no employment of the categories at all. Now the categories can only be applied *a priori* to intuition in so far as they grasp this form itself as a unitary representation. For categories are nothing else but forms of synthesizing into a unity. And by virtue of this, the application of the categories to all sensuous representations is also assured. For no sensible intuitions can be given *independently* of the forms of sensibility, which, in turn, are completely subordinate to the categories.

By reasoning in this way it is possible to maintain that the result Kant attains in section 26 on the basis of the mere fact of the givenness of unitary representations of space and time can be derived as a *necessary* condition of every employment of the categories. In accordance with this, the transcendental deduction no longer needs to be carried through in terms of those two steps of the proof which are characteristic of the second version.

In the same context, however, Kant also indicated the reasons for retaining the proof construction of the second edition: this proof makes use of the *synthetic method*, i.e., it proceeds on the basis of the fact that two doctrines of the *Critique* are initially developed independently of one another – the doctrine of the categories as functions of unity in self-consciousness and the doctrine of space and time as given representations. The second step of the proof according to the synthetic method has recourse to the results of the Aesthetic as to facts. If it were conducted according to analytical method, then the *necessity* of the forms of intuitions would first have to be justified. This would then be followed by an Aesthetic showing which forms we really have at our disposal. Only then could the deduction be completed. But Kant was of the opinion that this method 'did not have the clarity and facility' characteristic of the synthetic method.[10] And this is the reason which made him unable to consider Beck's proposed construction as an improvement. Kant always had the tendency to make his theory convincing by virtue of its theoretical consequences and, as far as possible, to reduce analysis of its foundations to a minimum. He was intent upon changing the entire method of philosophical instruction and upon securing the convictions of his age against the dogmatists and against skepticism. He did not foresee that through this pressure for application he would disillusion the best speculative minds among his students and drive them to other paths.

IV

The papers documenting Kant's reflections on the different methods for a transcendental deduction postdate the second edition of the *Critique* by almost ten years. Of course it is possible to show that all the ideas necessary for a deduction according to the analytical method had been already available to him when he published the first edition of the *Critique*. But this does not mean that he had in mind, as he composed the second edition, the advantages and disadvantages of a deduction according to one or the other method, and that he expressly chose the synthetic method on the basis of such a comparison. The text of the *Critique* provides no support for such an interpretation. Within the structure which Kant had already given his book, the advantages of a construction according to the synthetic method were in any case obvious. This construction allowed him to ground the two fundamental positions of critical philosophy, the sensible *a priori* and the active role of the understanding in knowledge, separately – and then to unite them by means of a single argument.

But there were still other reasons which induced him to argue the proof of the deduction synthetically and to divide it accordingly into two steps. Besides the task of proving the objective validity of the categories, Kant also assigned to the deduction the task of making intelligible the possibility of relating the understanding to sensibility.[11] This task must not be confused with the other of which Kant speaks in the first preface to the *Critique*, when he distinguishes the subjective from the objective side of the deduction (A XVI). There he says that the subjective deduction is an investigation of those cognitive faculties upon which the possibility of a functional knowledge by means of the understanding rests. Such an investigation strives for more than the explanation of possibility. It seeks to elucidate the whole apparatus of knowledge, if only in a summary. The explanation of possibility proposes merely to remove a difficulty which arises out of the problem of critical philosophy itself: it assumes pure categories and then declares, however, that these categories are originally and essentially related to sensible intuition. Given the accepted idea of an *a priori* concept, this relation seems mysterious and inconceivable. And thus it must be shown that the meaning of an *a priori* concept can be so stipulated that it refers necessarily to intuition. And it must further be shown how one can represent the fact that the given intuition essentially depends on such concepts. This explanation of possibility can also be given in another form. It has then to show that categories as well as intuition cannot even be thought independently of their relation to one another. Moreover, the demonstration of the necessity of a relation between them can provide an answer to the problematical question concerning the possibility of their relations.

It is well known that Kant sought in the second edition to avoid the problems of the so-called subjective deduction. But that does not mean that he neglected the demand for an explanation of the possibility of relating the categories to intuitions. To be sure, Kant used the same words to distinguish between the two questions involved in the proof of the validity of the categories: the objective deduction is a proof *that* the intuitions are subject to the categories, while both the subjective deduction and the explanation of possibility are investigations of *how* they do this. But they are still two distinct investigations. Adickes and Paton have overlooked this distinction. For this reason they were obliged to consider the second step of the proof to be something which it clearly was not: a subjective deduction. At the same time, however, one may very well read the *whole* deduction as an explanation of the possibility of relating the categories to intuition.

Such an explanation, like the proof of validity, requires two steps of explanation. First it must be shown what the nature of a category actually is, given the fact that it is always at the same time related to a synthesis of intuition. And it must then be shown that such categories can exercise synthetic functions in intuition itself. These two parts of the explanation can be given *at the same time* with the two steps of the validity proof, according to which categories in general are valid without limitation. One cannot blame Kant for not separating the two investigations. For it is easily shown that the proof of the validity of the categories *must* enter into the explanation of the possibility of their relation to intuition. At the only place where Kant separates the two investigations from one another, he was compelled to propose a proof of validity which fails to satisfy strict demands:[12] he has to *proceed* at this point *from the assumption* that we are in possession of synthetic *a priori* judgements concerning all objects of sensibility and that these judgements stand beyond all doubt in virtue of their employment in mathematical natural sciences. But this was the very presupposition which Hume called into question. And it is Kant's merit to have answered the radicalism of Hume's assault with a correspondingly radical founding of knowledge. For this reason, he ultimately did not incorporate into the *Critique* that form of a deduction which avoids every explanation of possibility. What we find in the second edition is a proof of the validity of the categories which is at one and the same time an explanation of the possibility of their relation to sensibility, a proof which avoids taking up the problems of an analysis of the cognitive faculties. And this is equally true of *both* steps of the deduction – not merely of its second part, which Adickes and Paton regard as a subjective deduction.

For an understanding of the text, both functions of the deduction must be borne in mind. But the proof of the validity of the categories always remains fundamental for a deduction. The second step of the proof, in section 26, overcomes the restriction still in effect in section 20. But considerable effort is required to recognize this second step. For it is not presented separately

from the second part of the explanation of possibility, and Kant himself adds
to the difficulty of understanding when he declares, with great emphasis, that
the unity in the representations of space and time can be none other than
that which is thought through the categories.[13] This statement, however, is
only an application of the result of section 20 and of the conclusion of
section 26. It contains neither a step of the proof nor a supplementary
explanation of possibility. It is simply misleading to give an application so
much stress that the actual conclusion of an argument is lost sight of.

                                      V

   But even with all these considerations it has still not been sufficiently
clarified why Kant never presented the actual structure of his proof in a
clearer manner or never even indicated it unequivocally. We have been
able to reconstruct it only from a text which constantly involves other
elements and at times gives them undue emphasis. A further reason can be
given for this circumstance which leads into a fundamental consideration
about the second version of the deduction and its relation to the first.
Already in section 16 Kant seems to suggest that he has completed the proof
of the deduction that all sensible representations are subject to the
categories. Kant's argument at this point makes use of an analysis of the
meaning expressed when I say that a representation is my representation.
The unity of apperception is the original definition of the meaning of
'belongs to me'. For this unity is indicated by the expressions 'I' or 'I think.'
All representations are therefore mine in so far as they are bound together
in the unity of self-consciousness. But Kant now extends the meaning of
'mine' further, namely, to include all representations which *can* be united in
self-consciousness. There are good grounds for this extension. For we do
actually call representations ours in virtue of the fact that we can become
conscious that they belong to us. And there is no better criterion for the
association of representations with a consciousness than the fact that it can
experience them as its own representations.
   This extension is equivocal nevertheless. For it leads easily to the more
extensive thesis that all representations which arise in the sensibility of a
being are to that degree also already potentially *conscious* representations.
Precisely because every consciousness presupposes, according to Kant, a
sensibility, one is tempted to describe this sensibility itself as 'his' sensibility
and then further assume that all given sensible representations must also to
that extent be 'his'. This would mean that they are to be defined as possible
contents of his consciousness. And, by proceeding in this way, the trans-

cendental deduction would be completed as a result of a simple semantic analysis of how the word 'mine' is used. For if all given representations are 'mine' in the sense indicated, then that means precisely that they can be taken up into the unity of consciousness in accordance with the categories.

That would be an astonishingly simple solution to the problem which Kant had announced as the *deepest* in the whole transcendental philosophy. Actually it rests, as is obvious, on a shift of meaning in the expression 'mine'. Sensibility is distinct from self-consciousness. And if something belongs to me only if I can take it up into consciousness, then as long as it is only *available* to be taken up into consciousness, it is not at all 'mine'; but only 'in relation to me'. It makes no difference whether this relation is public or private. If the limits of my consciousness' capacity to take something up preclude its entering my consciousness, it would also never become 'mine' in the strict sense of the word.

Kant was certainly aware of this distinction. At an important point he refrains from saying that there could be no intuition at all which would remain inaccessible to consciousness. He affirms only that this representation would then be 'nothing for me' (B 132). But it must also be noted that Kant did not seriously assume that there could be such representations – and not merely in virtue of the proof of the deduction, whether it be construed according to the analytical or the synthetic method. He did not see with sufficient clarity the homonymy of the word 'mine' in talk about 'my' representations. He was thus able to assert propositions which anticipate the result of the proof of the deduction and at the same time make the deduction dependent on the mere semantic analysis of the word 'mine': 'I am conscious of the self as identical with respect to the manifold of representations that are given to me in an intuition, because I call them one and all *my* representations, and so apprehend them as constituting *one* intuition. This amounts to saying that I am conscious *a priori* of a necessary synthesis of representations – to be entitled the original synthetic unity of apperception – under which all representations that are given to me must stand, . . .' (B 135.) In the first edition Kant made use of an even poorer argument in order to make the same conclusion convincing (A 120). In the second edition one can clearly recognize that he could not free himself from such arguments, but also that he does not rely upon them confidently. And for this reason he never expressly stated that the deduction already takes place in section 16. Instead he assures us that it *requires* both of the steps which we have analyzed (B 145). And they make no use at all of the semantic analysis of the word 'mine'.

One could show that it was no accident that Kant was upset by the ambiguity of this word. The difficulty inevitably occurs if one takes his doctrine of consciousness as a point of departure. Its distinctive features, however, cannot be examined here.

For the present discussion it is more important to see that this difficulty led to entirely different consequences in the two editions of the *Critique*. This difficulty is at least partly responsible for giving the second edition deduction an external form which is confusing and misleading. If we take the necessary pains, we can, nevertheless find an argument there which is convincing in the context of Kant's philosophy. In the first version, however, there is *no* proof which can dispense with the misleading argument from the double meaning of 'mine'. One can easily see this if one tries to rediscover in the first edition that thought which is indispensable in the second edition's division of the proof into two steps: the idea that the representations of space and time themselves presuppose a synthesis which includes everything that is given in them. To be sure, Kant took over this idea from the first edition, but at the same time he gave it an entirely different function. There it occurred only in the *introduction* to the proof (A 99, 101/2, 107). It seeks to clarify the fact that our cognitive process consists in something more than empirical powers and principles of combination which could only be investigated by association psychology. An *a priori* synthesis is also involved. Evidence of this is given in our representations of space and time, which cannot be understood without *a priori* synthesis. And this kind of synthesis leads to the philosophical question concerning the conditions of its possibility. By means of this argument Kant thus introduces into the first edition those principles in terms of which a transcendental deduction is to be given. In the second edition it has become an essential *part of the deduction* itself.

Thus there is a substantial difference between the proofs of the first and the second editions, and not merely in the manner of presentation, as Kant himself believed. We can understand why he himself was *unable* to see this distinction. For even in the second edition he did not yet altogether give up the inadequate argument that had been *absolutely* indispensable for the proof of the first edition. But as long as incompatible proof-strategies still continued to affect Kant's conception of the argument, he was not able to give an unequivocal presentation of the actual structure of the proof and to distinguish it clearly from the many related intentions which a transcendental deduction has to take into account at the same time. A careful stylistic analysis of the second edition reveals that Kant has modified his language in section 20 and that thereby he reaches for the first time the paths of the deduction which was to offer a cogent argument.

VI

We have noted previously that the proof of the second edition has the advantage over a possible analytical line of argument that it is better suited to the structure of the *Critique*. Now it has become apparent that, in comparison with the proof of the first edition, it also has the more significant advantage of being a formally correct proof. But these are not the only reasons for believing that Kant's thinking is more appropriately expressed in the second version of the deduction. An analysis of the proofs yields a far more general conclusion.

Now if it were only the structure of the book which recommends the proof of the second edition, one could, with Beck, suggest an alternative external shape for the *Critique* and thus a change in the form of its central argument. In point of fact, however, the second deduction is in complete agreement not only with the structure of the book but also with Kant's fundamental philosophical idea about the methodology of a philosophical system.[14]

Kant based this system on the unifying principle of the unity of self-consciousness. All its propositions must be deducible from this principle. But the method of this derivation does not consist in developing analytical implications from the concept of self-consciousness. Instead it specifies the presuppositions of the *possibility of the existence* of self-consciousness. By using this method, one can come to a knowledge of conditions which, although they are not already given in the structure of self-consciousness itself, must precisely *in virtue* of this structure be presupposed if a self-consciousness is to become actual.

This manner of argument is distinctively Kantian. It combines two propositions which may be regarded as the two formal tenets of the Kantian system: there must be a unifying principle in terms of which knowledge can be understood; yet this principle must not entail a monistic exclusion of all other principles; it must take into account the discovery of the essential difference in the roots of our knowledge and make possible a *raisonnement* which presupposes their underivable synthesis.

This method of argumentation is just as distinct from the faculty-psychology of empiricism as it is from those doctrines of the universal implications of the Ego which transformed Kant's position into that of speculative idealism. Empiricism lacked any principle of unity. The form of idealism which was historically so influential had no concept of an essential unity of originally *distinct* elements. Kant's transcendental deduction – *but only in its second version* – contains such a concept, although it is not fully developed. It proceeds, albeit with difficulty, according to a method of proof which is orientated by this concept.

If one succeeds in reaching a clear understanding of this method, one possesses the key to an understanding of Kant which is independent of his specific formulations. But what is more important, one will be able to understand those problems which remained insoluble for Kant as well as for his successors: the structure of consciousness, of moral knowledge, and of temporal experience.

It was only very late in his philosophical development that Kant worked out a balance between the two elements of such a method. First he realized the necessity of assuming distinct principles of knowledge whose interrelation is necessary. Later he discovered the unitary principle in terms of which such relations must be interpreted. Under the influence of this discovery, however, he maintained for more than a decade plans for a deduction which transcended the capacity of the unifying principle and which brought him into contradiction with his earlier discovery. Thus, for example, one can show that his *moral* philosophy was freed from more extensive claims of a deduction from self-consciousness and assumed its ultimate form only shortly before the appearance of the *Critique of Practical Reason*.[15] The change in the conception of a transcendental deduction corresponds fully to this development. And this correspondence is certainly not the weakest confirmation of the proposed interpretation.

Kant himself did not reach a clear understanding of the nature of these changes. And he withheld from his readers the clarity which he actually did possess for strategic reasons. Thus one cannot blame his successors if, unable to discover the coherence of his texts, they began to seek the spirit of critical philosophy in that conception of the nature of the philosophic system which Kant himself had just recently dismissed. In any case they were right in so far as the peculiar content of critical philosophy is only disclosed by autonomous philosophical effort. This task is still uncompleted today. But the solution of the engima involved in an interpretation of the transcendental deduction shows where this effort has to be directed.

[1] In this paper I shall discuss only the proof-structure of the Transcendental Deduction. An analysis of its premises and the problems involved in the application of its conclusion will be given in another paper.

[No such paper has appeared in English as yet, but cf. Henrich's monograph *Identität und Objektivität* (Heidelberg: Carl Winter Universitätsverlag, 1976). – *Ed.*]

[2] Cf., for instance, Norman Kemp Smith, *A Commentary on Kant's Critique of Pure Reason* (London: Macmillan, 1918), 289; and A. C. Ewing, *A Short Commentary on Kant's Critique of Pure Reason* (London: Methuen, 1938), 120.

[3] In recent English publications on the *Critique* one does not find a discussion of this problem. Bennett and Strawson are writing from a point of view which does not focus on it. Wolff is interested almost exclusively in the first edition of the Deduction.

[4] Erich Adickes, *Kants Kritik der reinen Vernunft* (Berlin, 1889), 139–40; Herbert James Paton, *Kant's Metaphysic of Experience* (London: Allen and Unwin, 1936), vol. 1, 501.

[5] Benno Erdmann, *Kants Kritizismus in der 1. and 2. Auflage der Kritik der reinen Vernunft* (Leipzig, 1878); Herman de Vleeschauwer, *La déduction transcendentale dans l'œuvre de Kant* (Paris, Antwerp and The Hague, 1934–7), vol. III, 24 ff.

[6] Cf. p. 160 of his translation, which shows also that, as a commentator, he could not find sense in Kant's text (cf. note 2 above).

[7] There was a thesis on the Transcendental Deduction by Friedrich Tenbruck (Marburg, 1944), never published, which came close to the conclusion of this section. Pietro Chiodi, *La Deduzione nell Opera di Kant* (Torino, 1961), 245 ff. makes an attempt to bring the problem of the 'how' (section 26) with that of the 'that' (section 20) into a necessary connection so that together they build one chain of arguments. But one cannot distinguish the two sections on the basis of these two problems. And moreover, Chiodi's account of Kant's intention is highly formal and abstract and cannot be expressed in the language of the Transcendental Deduction itself.

[8] *Erläuternder Auszug aus Kants kritischen Schriften*, vol. I (Riga, 1793); vol. III (Riga, 1796).

[9] *Kants gesammelte Schriften*, ed. Preussische Akademie der Wissenschaften, vol XII, letter to Tieftrunk 11 Dec. 1797, and the first sketch of this letter in vol. XIII, 467 ff.
Further evidence in vol. XVIII, reflections 6353 and 6358.

[10] *Kants gesammelte Schriften*, vol. XIII, 471.

[11] B 159: 'We have now to explain the possibility . . .'

[12] Cf. the note to the preface to Kant's *Metaphysical Foundations of Science (Kants gesammelte Schriften*, vol. IV, 474 ff.). Like the *Prolegomena* it starts from science as an indubitable fact, which is, according to the *Prolegomena*, legitimate only within an introduction into the *Critique*. The differences between the *Prolegomena* and the note may be ignored here.

[13] B 161: 'This synthetic unity can be no other than . . .'

[14] D. Henrich, 'Zu Kants Begriff der Philosophie,' in *Kritik und Metaphysik, Festschrift für Heinz Heimsoeth* (Berlin, 1966), 40 ff.

[15] This is shown in various articles of mine, among them: 'Der Begriff der sittlichen Einsicht und Kants Lehre vom Faktum der Vernunft' in: *Die Gegenwart der Griechen im neueren Denken* (Tübingen, 1960), 77 ff.; and 'Das Problem der Grundlegung der Ethik bei Kant und im spekulativen Idealismus' in *Sein und Ethos* (Mainz, 1963), 350 ff.

IV

IMAGINATION AND PERCEPTION

P. F. STRAWSON

*Psychologists have hitherto failed to
realize that imagination is a necessary
ingredient of perception itself.*[1]

I

THE uses, and applications, of the terms 'image', 'imagine', 'imagination', 'imaginative', and so forth make up a very diverse and scattered family. Even this image of a family seems too definite. It would be a matter of more than difficulty exactly to identify and list the family's members, let alone establish their relationships of parenthood and cousinhood. But we can at least point to different areas of association in each of which some members of this group of terms ordinarily find employment. Here are three such areas: (1) the area in which imagination is linked with *image* and image is understood as *mental image* – a picture in the mind's eye or (perhaps) a tune running through one's head; (2) the area in which imagination is associated with invention, and also (sometimes) with originality or insight or felicitous or revealing or striking departure from routine; (3) the area in which imagination is linked with false belief, delusion, mistaken memory, or misperception. My primary concern here is not with any of these three areas of association, though I shall refer to them all, and especially to the first. My primary topic is Kant's use of the term 'imagination', in the *Critique of Pure Reason*, in connection with perceptual recognition – a use which may appear something of an outsider, but nevertheless has claims to affinity which are worth considering. I shall refer also to Hume and to Wittgenstein. My paper in general belongs to the species *loosely ruminative* and *comparative-historical* rather than to the species *strictly argumentative* or *systematic-analytical*.

First published in Foster and Swanson (eds.), *Experience and Theory*. Amherst, Mass., and London: University of Massachusetts Press and Duckworth, 1971, pp. 31–54; reprinted in P. F. Strawson, *Freedom and Resentment and Other Essays*. London: Methuen, 1974, pp. 45–65. Reprinted by permission of the author and Associated Book Publishers Ltd.

II

Sometimes Kant used the term 'imagination' and its cognates in what is apparently a very ordinary and familiar way; as when, for example, he seems to contrast our imagining something with our having knowledge or experience of what is actually the case. Thus in a note in the 'Refutation of Idealism' he writes: 'It does not follow that every intuitive representation of outer things involves the existence of these things, for their representation can very well be the product *merely of the imagination* (as in dreams and delusions). . . . Whether this or that supposed experience be not *purely imaginary*, must be ascertained from its special determinations, and through its congruence with the criteria of all real experience.'[2] Sometimes, however, indeed more frequently, his use of the term seems to differ strikingly from any ordinary and familiar use of it, so that we are inclined to say he must be using it in a technical or specialized way of his own. Suppose, for example, that I notice a strange dog in the garden, and observe its movements for a while; and perhaps also notice, a few minutes later, that it is still there. We should not ordinarily say that this account of a small and uninteresting part of my history included the report of any exercise of the imagination on my part. Yet, in Kant's apparently technical use of the term, any adequate analysis of such a situation would accord a central role to imagination, or to some faculty entitled 'imagination'.

In both these respects there is a resemblance between Kant and Hume. That is to say, Hume, like Kant, sometimes makes an apparently ordinary use of the term (as when he is discussing the differences between imagination and memory) and sometimes makes an apparently technical use of it; and the latter use is such that he, too, like Kant, would say that imagination enters essentially into the analysis of the very ordinary situation I described a moment ago. It may be instructive to see how far this resemblance goes.

Let us return to our simple situation. Both Hume and Kant would say (a) that my recognizing the strange dog I see as a dog at all owes something to the imagination; and (b) that my taking what I continuously, or interruptedly, observe to be the same object, the same dog, throughout, also owes something to the imagination. By both philosophers imagination is conceived as a connecting or uniting power which operates in two dimensions. In one dimension, (a), it connects perceptions of different objects of the same kind; in the other dimension, (b), it connects different perceptions of the same object of a given kind. It is the instrument of our perceptual appreciation both of kind-identity and of individual-identity, both of concept-identity and of object-identity. The two dimensions or varieties of connecting power are, doubtless, not independent of each other, but they

can, to some extent, be handled separately. I begin by referring briefly to
(a); then I treat more fully of (b); and then return in section IV, below, to
(a).

Kant's doctrine (or part of it) on (a) is sketched in the chapter on
schematism, and Hume's in the chapter in the *Treatise* called 'Of abstract
ideas'. Kant declares the schema to be a product of, and also a rule for, the
imagination, in accordance with which, and by means of which alone, the
imagination can connect the particular image or the particular object with
the general concept under which it falls. Hume speaks, in his usual way, of
the resemblance of particular ideas being the foundation of a customary
association both among the resemblant particular ideas themselves and
between them and the 'annex'd' general term; so that the imagination is, or
may be, ready with an appropriate response whenever it gets a cue, as it
were, from anywhere in the associative network. How the mechanism is
supposed exactly to work is not very clear either in the case of Hume or in
that of Kant. But the obscurity of this very point is something which both
authors emphasize themselves, in sentences which show a quite striking
parallelism. Thus Kant says of schematism that it is 'an art concealed in the
depths of the human soul, whose real modes of activity nature is hardly
likely ever to allow us to discover, and to have open to our gaze'.[3] And
Hume, speaking of the imagination's readiness with appropriate particular
ideas, describes it as a 'kind of magical faculty in the soul which, though it be
always most perfect in the greatest geniuses, and is properly what we call a
genius, is, however, inexplicable by the utmost efforts of human understanding'.[4] Imagination, then, in so far as its operations are relevant to the
application of the same general concept in a variety of different cases, is a
concealed art of the soul, a magical faculty, something we shall never fully
understand.

Let us turn now to (b), to the matter of different phases of experience
being related to the same particular object of some general type. In both
authors this question is absorbed into a larger one, though the larger
question is somewhat differently conceived in each of them. The main
relevant passages here are, in the *Critique*, the section on 'Transcendental
Deduction' and, in the *Treatise*, the chapter 'Scepticism with Regard to the
Senses'. Let us begin with Hume.

Hume makes a threefold distinction between sense (or the senses),
reason (or understanding), and imagination. His famous question about the
causes which induce us to believe in the existence of body resolves itself into
the question to which of these faculties, or to what combination of them, we
should ascribe this belief, that is, the belief in the *continued* and *distinct*
existence of bodies. Certainly, he says, not to the senses alone and unassisted.

For 'when the mind looks further than what immediately appears to it, its conclusions can never be put to the account of the senses';[5] and the mind certainly 'looks further' than this, both in respect of the belief in the *continued* existence of objects when we are no longer, as we say, perceiving them, and in respect of the obviously connected belief in the *distinctness* of their existence from that of our perceptions of them. Equally certainly, he says, we cannot attribute these beliefs to Reason, that is to reasoning based on perceptions. For the only kind of reasoning that can be in question here is reasoning based on experience of constant conjunction, or causal reasoning. But whether we conceive of objects as the same in kind as perceptions or as different in kind from perceptions, it remains true that 'no beings are ever present to the mind but perceptions'[6] and all perceptions which are present to the mind are present to the mind;[7] hence it is equally certain that we can never observe a constant conjunction either between perceptions present to the mind and perceptions not present to the mind or between perceptions on the one hand and objects different in kind from perceptions on the other.

The belief in question, then, must be ascribed to the Imagination – or, more exactly, to the 'concurrence' of some of the qualities of our impressions with some of the qualities of the imagination. And here Hume launches into that famous account of the operations of imagination which, on account of its perverse ingenuity, can scarcely fail to command admiration both in the original and the modern senses of the word. It runs roughly as follows: imagination engenders so strong a propensity to confound the similarity of temporally separated and hence non-identical perceptions with strict identity through time that, in defiance of sense and reason combined, we feign, and believe in, a continued existence of perceptions where there is patently no such thing; and so strong is the hold of this belief that, when the discrepancy is pointed out, the imagination can still find an ally in certain philosophers who try, though vainly, to satisfy reason and imagination at the same time by conceiving of objects as different in kind from perceptions and ascribing continued existence to the former and interrupted existence only to the latter.

When we turn from Hume to Kant, it is probably the divergencies rather than the parallels which we find most striking in this case – at least at first. And perhaps we can come at these by considering a *simpliste* criticism of Hume. For Hume's account is full of holes. One of the most obvious relates to his bland assertion that the unreflective, as opposed to the philosophers, take the objects of perception to be of the same species as perceptions of those objects; so that the problem of accounting for the belief, in its *vulgar* form, in the continued existence of objects is the problem of accounting for a belief which reason shows to be ungrounded and ungroundable, namely a belief in the existence of perceptions which nobody has. Of course it is quite false that the vulgar make any such identification and hence quite false that they hold any such belief as Hume presumes to account for. The vulgar *distinguish*, naturally and unreflectively, between their seeings and hearings (perceivings) of objects and the objects they see and hear, and hence have no difficulty in reconciling the interruptedness of the former with the continuance of existence of the latter. Indeed these distinctions and beliefs are built into the very vocabulary of their perception-reports, into the concepts they employ, the meanings of the things they say, in giving (unsophisticated) accounts of their hearings and seeings of things. So Hume's problem does not really exist and his solution to it is otiose.

I think Kant would regard these criticisms as just, but would deny that there was therefore no problem at all for the philosopher. That is, he would agree that the problem was not, as Hume conceived it, that of accounting, on the basis of the character of our perceptual experience, for certain beliefs (beliefs in the continued and distinct existence of bodies). For he would agree that it would be impossible to give accurate, plain reports of our perceptual experience which did not already incorporate those beliefs. The beliefs form an essential part of the conceptual framework which has to be employed to give a candid and veridical description of our perceptual experience. But this does not mean that there is no question to be asked. Hume starts his investigation, as it were, too late; with perceptual experience already established in the character it has, he leaves himself no room for any such question as he wishes to ask. But we ought to ask, not how it can be that on the basis of perceptual experience as it is, we come to have the beliefs in question, but *how it is* that perceptual experience is already such as to embody the beliefs in question; or, perhaps better, *what* it is for perceptual experience to be such as to embody the beliefs in question.

I do not want to invoke more of the complex apparatus of the critical philosophy than is necessary to bring out the parallels with Hume that lie below or behind or beside the divergencies. We know that Kant thought that perceptual experience did not just *happen* to have the general character it has, but *had* to have at least something like this character, if experience (that is the temporally extended experience of a self-conscious being) was to be possible at all. Just now we are not so much concerned with the soundness of this view as with the question of what he thought was *involved* in perceptual experience having this character. One of the things he certainly thought was involved is this: 'A combination of them [perceptions or representations], such as they cannot have in sense itself, is demanded.'[8] And this 'such as they cannot have in sense itself' arouses at least a faint echo of Hume's view that sense itself could never give rise to the *opinion* of the continued and distinct existence of body. The reason Hume gives for this view, it will be recalled, is that in embracing such an *opinion*, 'the mind looks further than what immediately appears to it'. Now could Kant have a *similar* reason for holding that, for the use of concepts of relatively permanent bodies (that is for perceptual experience to have the character it does have), a combination such as perceptions cannot have in sense itself is demanded?

I think he could have. For even when Hume is submitted to the sort of correction I sketched above, there is *something* right about the phrase of his I have just quoted. When I naïvely report what I see at a moment (say, as a tree or a dog), my mind or my report certainly 'looks further' than *something* – not, usually, than 'what immediately appears to me' (tree or dog), but certainly further than the merely subjective side of the event of its immediately appearing to me. Of a fleeting perception, a subjective event, I give a description involving the mention of something not fleeting at all, but lasting, not a subjective event at all, but a distinct object. It is clear, *contra* Hume, not only that I *do* do this, but that I *must* do it in order to give a natural and unforced account of my perceptions. Still, there arises the question of what is necessarily involved in this being the case. The uninformative beginnings of an answer consist in saying that one thing necessarily involved is our possession and application of concepts of a certain kind, namely concepts of distinct and enduring objects. But now, as both Kant and Hume emphasize, the whole course of our experience of the world consists of relatively transient and changing perceptions. (The changes, and hence the transience, may be due to changes in the scene or in our orientation, broadly understood, towards the scene.) It seems reasonable to suppose that there would be no question of applying concepts of the kind in question unless those concepts served in a certain way to *link* or *combine* different perceptions – unless, specifically, they could, and sometimes did, serve to link *different* perceptions as perceptions of the *same* object. Here, then, is one aspect of combination, as Kant uses the word, and just the aspect we are now concerned with. Combination, in this sense, is *demanded*. We could not count any transient perception as a perception of an enduring object of some kind unless we were prepared to count, and did count, some transient perceptions as, though different perceptions, perceptions of the same object of such a kind. The concepts in question could get no grip at all unless different perceptions were sometimes in this way combined by them. And when Kant says that this sort of combination of perceptions is such as they (the perceptions) cannot have in sense itself, we may perhaps take him to be making *at least* the two following unexceptionable, because tautological, points:

(1) that this sort of combination is dependent on the possession and application of this sort of concept, that is, that if we did not *conceptualize* our sensory intake in this sort of way, then our sensory impressions would not be *combined* in this sort of way;

(2) that distinguishable perceptions combined in this way, whether they are temporally continuous (as when we see an object move or change colour) or temporally separated (as when we see an object again after an interval), really are distinguishable, that is different, perceptions.

Of course, in saying that we find these two unexceptionable points in Kant's Hume-echoing dictum about combination, I am not for a moment suggesting that this account covers all that Kant means by combination; only that it may reasonably be taken to be included in what Kant means.

But now how does imagination come into the picture, that is into Kant's picture? Kant's problem, as we have seen, is not the same as Hume's; so he has no call to invoke imagination to do the job for which Hume invokes it, that is the job of supplementing actual perceptions with strictly imaginary perceptions which nobody has, which there is no reason to believe in the existence of and every reason not to believe in the existence of, but which we nevertheless *do* believe in the existence of as a condition of believing in the existence of body at all. This is not how imagination can come into Kant's picture. But certainly imagination *does* come into his picture; and the question is whether we can give any intelligible account of its place there. I think we can give some sort of account, though doubtless one that leaves out much that is mysterious in Kant and characteristic of him.

To do this we must strengthen our pressure at a point already touched on. We have seen that there would be no question of counting any transient perception as a perception of an enduring and distinct object unless we were prepared or ready to count some different perceptions as perceptions of one and the same enduring and distinct object. The thought of *other* actual or possible perceptions as related in this way to the *present* perception has thus a peculiarly intimate relation to our counting or taking – to our ability to count or take – this present perception as the perception of such an object. This is not of course to say that even when, for example, we perceive and recognize (reidentify as the object it is) a familiar particular object, there need occur anything which we could count as the experience of actually recalling any particular past perception of that object. (It is not in *this* way, either, that imagination comes into the picture.) Indeed the more familiar the object, the less likely any such experience is. Still, in a way, we can say in such a case that the past perceptions are *alive* in the present perception. For it would not be just the perception it is but for them. Nor is this just a matter of an external, causal relation. Compare seeing a face you *think* you know, but cannot associate with any previous encounter, with seeing a face you *know* you know and can very well so associate, even though there does not, as you see it, occur any particular *episode* of recalling any particular previous encounter. The comparison will show why I say that the past perceptions are, in the latter case, not merely causally operative, but alive in the present perception.

Of course when you first see a new, an unfamiliar thing of a familiar kind, there is no question of past perceptions of *that* thing being alive in the present perception. Still, one might say, to take it, to see it, as a thing of that kind is implicitly to have the thought of other possible perceptions related to your actual perception as perceptions of the same object. To see it as a dog, silent and stationary, is to see it as a possible mover and barker, even though you give yourself no actual images of it as moving and barking; though, again, you might do so if, say, you were particularly timid, if, as we say, your imagination was particularly active or particularly stimulated by the sight. Again, as you continue to observe it, it is not just a dog, with such and such characteristics, but *the* dog, the object of your recent observation, that you see, and see it as.

It seems, then, not too much to say that the actual occurrent perception of an enduring object as an object of a certain kind, or as a particular object of that kind, is, as it were, soaked with or animated by, or infused with – the metaphors are *à choix* – the thought of other past or possible perceptions of the same object. Let us speak of past and merely possible perceptions alike as 'non-actual' perceptions. Now the imagination, in one of its aspects – the first I mentioned in this paper – is the image-producing faculty, the faculty, we may say, of producing actual representatives (in the shape of images) of non-actual perceptions. I have argued that an actual perception of the kind we are concerned with owes its character essentially to that internal link, of which we find it so difficult to give any but a metaphorical description, with other past or possible, but in any case non-actual, perceptions. Non-actual perceptions are in a sense represented in, alive in, the present perception; just as they are represented, by images, in the image-producing activity of the imagination. May we not, then, find a kinship between the capacity for this latter kind of exercise of the imagination and the capacity which is exercised in actual perception of the kind we are concerned with? Kant, at least, is prepared to register his sense of such a kinship by extending the title of 'imagination' to cover both capacities; by speaking of imagination as 'a necessary ingredient of perception itself'.

III

Suppose we so understand – or understand as including at least so much – the Kantian idea of the synthesis of imagination. The connection of the idea, so understood, with the application of concepts of objects is already clear. Can we also explain the introduction of the qualification 'transcendental'? If we bear in mind the opposition between 'transcendental' and 'empirical', I think we can put two glosses on 'transcendental' here, both with a common root. First, then, we must remember the distinction between what Kant thought necessary to the possibility of any experience and what he thought merely contingently true of experience as we actually enjoy it. There is, in this sense, no necessity about our employment of the particular sets of empirical concepts we do employ, for example the concepts of elephant or ink bottle. All that is necessary is that we should employ some empirical concepts or other which exemplify, or give a footing to, those very abstractly conceived items, the categories, or concepts of an object in general. Synthesis, then, or the kind of exercise of the imagination (in Kant's extended sense) which is involved in perception of objects *as* objects, is empirical in one aspect and transcendental in another: it is empirical (that is non-necessary) in so far as it happens to consist in the application of this or that particular empirical concept (elephant or ink bottle); transcendental (that is necessary) in so far as the application of such concepts represents, though in a form which is quite contingent, the utterly general requirements of a possible experience.

The second, connected, gloss we can put upon 'transcendental' can be brought out by comparison, once more, with Hume. Hume seems to think of the operations of imagination as something superadded to actual occurrent perceptions, the latter having a quite determinate character independent of and unaffected by the imagination's operations (though, of course, our *beliefs* are not unaffected by those operations). The Kantian synthesis, on the other hand, however conceived, is something necessarily involved in, a necessary condition of, actual occurrent reportable perceptions having the character they do have. So it may be called 'transcendental' in contrast with any process, for example any ordinary associative process, which presupposes a basis of actual, occurrent, reportable perceptions.

IV

In so far as we have supplied anything like an explanation or justification of Kant's apparently technical use of 'imagination', we have done so by suggesting that the recognition of an enduring object of a certain kind *as* an object of that kind, or as a certain particular object of that kind, involves a certain sort of connection with other non-actual perceptions. It involves other past (and hence non-actual) perceptions, or the thought of other possible (and hence non-actual) perceptions, of the *same* object being somehow alive in the present perception. The question arises whether we can stretch things a little further still to explain or justify the apparently technical use of 'imagination' in connection with our power to recognize *different* (and sometimes very different) particular objects as falling under the same general concept.

We can begin by making the platitudinous point that the possession of at least a fair measure of this ability, in the case, say, of the concept of a tree, is at least a test of our knowing what a tree is, of our possessing the concept of a tree. And we can progress from this to another point, both less platitudinous and more secure: namely, that it would be unintelligible to say of someone that whereas he could recognize *this* particular object as a tree, he could not recognize any other trees as trees.[9] So it would not make sense to say, in the case of a particular momentary perception, that he who had it recognized what he saw as a tree unless we were prepared also to ascribe to him the power of recognizing other things as well as trees. Now, how are we to regard this power or potentiality as related to his momentary perception? Is it just something external to it, or superadded to it, just an extra qualification he must possess, as it were, if his momentary perception is to count as a case of tree-recognition? This picture of the relation seems wrong. But if we say it *is* wrong, if we say that the character of the momentary perception itself depends on the connection with this general power, then have we not in this case too the same sort of link between actual and non-actual perceptions (now of *other* things) as we had in the previously discussed case between actual and non-actual perceptions (then of the *same* thing)? But if so then we have another reason, similar to the first reason though not the same as it, for saying that imagination, in an extended sense of the word, is involved in the recognition of such a thing as the sort of thing it is. Once more, this is not a matter of supposing that we give ourselves actual images, either of other trees perceived in the past or of wholly imaginary trees not perceived at all, whenever, in an actual momentary perception, we recognize something as a tree. It is not in this way, that is, by being represented by actual images, that non-actual (past or possible) perceptions enter into actual perception. They enter, rather, in that elusive way of which we have tried to give an account. But may we not here again, for this very reason, find a kinship between perceptual recognition (of an object as of a certain kind) and the more narrowly conceived exercise of the imagination –

enough of a kinship, perhaps, to give some basis for Kant's extended use of the term 'imagination' in this connection too, and perhaps, this time, for Hume's as well?

V

It does not, of course, matter very much whether we come down in favour of, or against, this extended or technical application of the term 'imagination'. What matters is whether, in looking into possible reasons or justifications for it, we find that any light is shed on the notion of perceptual recognition. And here I want to summon a third witness. The third witness is Wittgenstein. I consider his evidence, first, in this section, without any reference to any explicit use, by him, of the term 'imagination'; then, in the next, I refer to some of his own uses of terms of this family.

On page 212 of the *Investigations* Wittgenstein says: 'We find certain things about seeing puzzling, because we do not find the whole business of seeing puzzling enough.' This comes nearly at the end of those twenty pages or so which he devotes to the discussion of *seeing as*, of aspects and changes of aspect. Nearly all the examples he considers, as far as visual experience is concerned, are of pictures, diagrams, or signs, which can present different aspects, can be seen now as one thing, now as another. He is particularly impressed by the case where they undergo a change of aspects under one's very eyes, as it were, the case where one is suddenly struck by a new aspect. What, I think, he finds particularly impressive about this case is the very obviously *momentary* or *instantaneous* character of the being struck by the new aspect. Why does this impress him so much? Well, to see an aspect, in this sense, of a thing is, in part, to *think* of it in a certain way, to be disposed to *treat* it in a certain way, to give certain sorts of explanations or accounts of what you see, in general to *behave* in certain ways. But, then, how, he asks, in the case of seeing an aspect, is this thinking of the thing in a certain way related to the *instantaneous* experience? We could perhaps imagine someone able to *treat* a picture in a certain way, painstakingly to *interpret* it in that way, without *seeing* the relevant aspect, without seeing it *as* what he was treating it as, at all.[10] But this does not help us with the case of the instantaneous experience. It would be quite wrong to speak of this case as if there were merely an external relation, inductively established, between the thought, the interpretation, and the visual experience: to say, for example, that 'I see the *x* as a *y*' means 'I have a particular visual experience which I have found that I always have when I interpret the *x* as a *y*.'[11] So Wittgenstein casts around for ways of expressing himself which will hit off the relation.

Thus we have: 'The flashing of an aspect on us seems half visual experience, half thought';[12] or again, of a different case, 'Is it a case of both seeing *and* thinking? or an amalgam of the two, as I should almost like to say?';[13] or again, of yet another, 'It is almost as if "seeing the sign in this context" [under this aspect] were an echo of a thought. "The echo of a thought in sight" – one would like to say.'[14]

Beside Wittgenstein's metaphor of 'the echo of the thought in sight' we might put others: the visual experience is *irradiated* by, or *infused* with, the concept; or it becomes *soaked* with the concept.

Wittgenstein talks mainly of pictures or diagrams. But we must all have had experiences like the following: I am looking towards a yellow flowering bush against a stone wall, but I see it as yellow chalk marks scrawled on the wall. Then the aspect changes and I see it normally, that is I see it as a yellow flowering bush against the wall. On the next day, however, I see it normally, that is I see it as a yellow flowering bush against the wall, all the time. Some persons, perhaps with better eyesight, might never have seen it as anything else, might always *see it as* this. No doubt it is only against the background of some such experience of change of aspects, or of the thought of its possibility, that it is quite natural and non-misleading to speak, in connection with ordinary perception, of *seeing* objects *as* the objects they are. But this does not make it incorrect or false to do so generally.[15] Wittgenstein was perhaps *over*-impressed by the cases where we are *suddenly* struck by something – be it a classical change of figure-aspects or the sudden recognition of a face or the sudden appearance of an object, as when an ordinary rabbit bursts into view in the landscape and captures our attention.[16] Though there clearly are distinctions between cases, there are also continuities. There is no reason for making a sharp conceptual cleavage between the cases of a sudden irruption – whether of an aspect or an object – and others. We can allow that there are cases where visual experience is suddenly irradiated by a concept and cases where it is more or less steadily soaked with the concept. I quote once more: 'We find certain things about seeing puzzling because we do not find the whole business of seeing puzzling enough.' Perhaps we should fail less in this respect if we see that the striking case of the *change* of aspects merely dramatizes for us a feature (namely seeing as) which is present in perception in general.

Now how do we bring this to bear on Kant? Well, there is a point of analogy and a point of difference. The thought is echoed in the sight, the concept is alive in the perception. But when Wittgenstein speaks of *seeing as* as involving thinking-of-as, as involving the thought or the concept, he has in mind primarily a disposition to behave in certain ways, to treat or describe what you see in certain ways – such a disposition itself presupposing (in a favourite phrase) the mastery of a technique. This is the *criterion* of the visual experience, the means by which someone other than the subject of it must tell what it is. This, taking us on to familiar Wittgensteinian ground, gives us indeed a peculiarly intimate link between the momentary perception and something else; but the 'something else' is behaviour, and so the upshot seems remote from the peculiarly intimate link we laboured to establish in connection with Kant's use of the term 'imagination': the link between the actual present perception of the object and other past or possible perceptions of the same object or of other objects of the same kind. But is it really so remote? Wittgenstein's special preoccupations pull him to the behavioural side of things, to which Kant pays little or no attention. But we can no more think of the behavioural dispositions as merely externally related to *other* perceptions than we can think of them as merely externally related to the present perception. Thus the relevant behaviour in reporting an aspect may be to point to *other* objects of *perception*.[17] Or in the case of seeing a real, as opposed to a picture-object, as a such-and-such, the behavioural disposition includes, or entails, a readiness for, or expectancy of, other perceptions, of a certain character, of the same object.

Sometimes this aspect of the matter – the internal link between the present and other past or possible perceptions – comes to the fore in Wittgenstein's own account. Thus, of the case of sudden recognition of a particular object, an old acquaintance, he writes: 'I meet someone whom I have not seen for years; I see him clearly, but fail to know him. Suddenly I know him, *I see the old face in the altered one*.'[18] Again, he says of the dawning of an aspect: 'What I perceive in the dawning of an aspect is . . . an internal relation between it [the object] and other objects.'[19]

VI

I have mentioned the fact that there are points in these pages at which Wittgenstein himself invokes the notions of imagination and of an image. I shall discuss these points now. He first invokes these notions in connection with the drawing of a triangle, a right-angled triangle with the hypotenuse downmost and the right angle upmost. 'This triangle', he says, 'can be seen as a triangular hole, as a solid, as a geometrical drawing; as standing on its base, as hanging from its apex; as a mountain, as a wedge, as an arrow or pointer, as an overturned object which is meant to stand on the shorter side of the right angle, as a half-parallelogram, and as various other things.'[20] Later he reverts to this example and says: 'The aspects of the triangle: it is as if an *image* came into contact, and for a time remained in contact, with the visual impression.'[21] He contrasts some of the triangle-aspects in this respect with the aspects of some other of his examples; and a little later he says: 'It is possible to take the duck-rabbit simply for the picture of a rabbit, the double cross simply for the picture of a black cross, but not to take the bare triangular figure for the picture of an object that has fallen over. To see this aspect of the triangle demands *imagination* [*Vorstellungskraft*].'[22] But later still he says something more general about seeing aspects. 'The concept of an aspect is akin to the concept of an image. In other words: the concept "I am now seeing it as . . . " is akin to "I am now having *this* image".'[23] Immediately afterwards he says: 'Doesn't it take imagination [*Phantasie*] to hear something as a variation on a particular theme? And yet one is perceiving something in so hearing it.'[24] Again on this page he says generally that seeing an aspect and imagining are alike subject to the will.

It is clear that in these references to imagination and to images Wittgenstein is doing at least two things. On the one hand he is *contrasting* the seeing of certain aspects with the seeing of others, and saying *of some only* that they require imagination; and, further, that some of these are cases in which an image is, as it were, in contact with the visual impression. On the other hand he is saying that there is a *general* kinship between the seeing of aspects and the having of images; though the only respect of kinship he mentions is that both are subject to the will. Perhaps we can make something of both of these.

As regards the first thing he is doing, the contrast he is making, cannot we find an analogy here with a whole host of situations in which there is some sort of departure from the immediately obvious or familiar or mundane or established or superficial or literal way of taking things; situations in which there is some sort of innovation or extravagance or figure or trope or stretch of the mind or new illumination or invention? Thus, beginning from such simplicities as seeing a cloud as a camel or a formation of stalagmites as a dragon, or a small child at a picnic seeing a tree stump as a table, we may move on to very diverse things: to the first application of the word 'astringent' to a remark or to someone's personality; to Wellington at Salamanca saying 'Now we have them' and seeing the future course of the battle in an injudicious movement of the enemy; to the sensitive observer of a personal situation seeing that situation as one of humiliation for one party and triumph for another; to a natural (or even a social) scientist seeing a pattern in phenomena which has never been seen before and introducing, as we say, new concepts to express his insight; to anyone seeing Keble College, Oxford, or the University Museum or Balliol Chapel as their architects meant them to be seen; to Blake seeing eternity in a grain of sand and heaven in a wild flower. And so on. In connection with any item in this rather wild list the words 'imaginative' and 'imagination' are appropriate, though only to some of them is the idea of an image coming into contact with an impression appropriate. But we must remember that what is obvious and familiar, and what is not, is, at least to a large extent, a matter of training and experience and cultural background. So it may be, in this sense, imaginative of Eliot to see the river as a strong, brown god, but less so of the members of a tribe who believe in river-gods. It may, in this sense, call for imagination on my part to see or hear something as a variation on a particular theme, but not on the part of a historian of architecture or a trained musician. What is fairly called exercise of imagination for one person or age group or generation or society may be merest routine for another. To say this is not, of course, in any way to question the propriety of using the term 'imagination' to mark a *contrast*, in any particular case, with routine perception in the application of a concept. It is simply to draw attention to the kind, or kinds, of contrast that are in question and in doing so to stress resemblances and *continuities* between contrasted cases. It should not take much effort to see the resemblances and continuities as at least as striking as the differences and so to sympathize with that imaginative employment of the term 'imagination' which leads both Hume and Kant to cast the faculty for the role of chief agent in the exercise of the power of concept-application, in general, over a variety of cases; to see why Hume described it as a 'magical faculty' which is 'most perfect in the greatest geniuses' and is 'properly what we call a genius'.

So we find a continuity between one aspect of Wittgenstein's use of the term and one aspect of Hume's and Kant's. What of the other aspect of Wittgenstein's use, where he finds a kinship, in *all* cases, between seeing an aspect and having an image? Well, let us consider the character of Wittgenstein's examples. Some are examples of what might be called essentially ambiguous figures, like the duck-rabbit or the double cross. Others are, as it were, very thin and schematic, like the cube-picture or the triangle. If we attend to the essentially ambiguous figures, it is clear that imagination in the sense just discussed would not normally be said to be required in order to see either aspect of either. Both aspects of each are entirely natural and routine, only they compete with each other in a way which is not usual in the case of ordinary objects. We can switch more or less easily from one aspect to another as we cannot normally do with ordinary objects of perception. But we might sometimes switch with similar ease in what on the face of it are ordinary cases: thus, standing at the right distance from my yellow flowering bush, I can switch from seeing it as such to seeing it as yellow chalk marks scrawled on the wall. So if the affinity between seeing aspects and having images is simply a matter of subjection to the will, and if subjection to the will is thought of in this way as ease of switching, then the affinity is present in this case as in the case of the visually ambiguous figures.

But is the general affinity between seeing aspects and having images simply a matter of subjection to the will? One may point out that the subjection of *seeing as* to the will is by no means absolute or universal. And it may be replied that the same is true of having images. One may be haunted or tortured by images, whether of recall or foreboding, from which one vainly seeks distraction but cannot dismiss, or escape the return of, if dismissed; or, alternatively, one may fail to picture something in one's mind when one tries.[25] So at least a *parallel* between *seeing as* and having images, in respect of subjection to the will, continues to hold.

But surely one may ask whether there is not a deeper affinity between *seeing as* and having an image, one which goes beyond this matter of subjection to the will, and can be found in general between perception and imaging. And surely there is. It has already been expressed in saying that the thought (or, as Kant might prefer, the concept) is alive in the perception just as it is in the image. The thought of something as an x or as a particular x is alive in the perception of it as an x or as a particular x just as the thought of an x or a particular x is alive in the having of an image of an x or a particular x. This is what is now sometimes expressed in speaking of the *intentionality* of perception, as of imaging.[26] But the idea is older than *this* application of that terminology, for the idea is in Kant.

Of course it is essential to the affinity that the having of an image, like perceiving, is more than just having a thought; and that the more that it is is what justifies us in speaking of an image as an actual representative of a non-actual perception and justifies Hume (for all the danger of it) in speaking of images as faint copies of impressions. As for the differences between them both in intrinsic character and in external, causal relations, there is perhaps no need to stress them there.

VII

I began this paper by mentioning three areas of association in which the term 'imagination' and its cognates find employment: in connection with *images*, in connection with *innovation* or *invention*, and in connection with *mistakes*, including perceptual mistakes. I have referred to the first two areas of use, but not, so far, to the last. But perhaps, it is worth glancing briefly at the quite common use of 'imagine' and 'imagination' in connection with the *seeing as* of perceptual mistakes. Suppose that when I see the yellow flowering bush as yellow chalk marks on the wall, I actually take what I see *to be* yellow chalk marks on the wall – as I may well do once, though probably not when I have the experience again. In such a case, as opposed to that of *seeing as* without *taking as*, it would be natural and correct to say: 'For a moment I imagined what I saw to be yellow chalk marks on the wall; then I looked again and saw it was a yellow flowering bush against the wall.'

Now it would be easy, and reasonable, to explain this 'mistake' use of 'imagine' by taking some other use or uses as primary and representing this use as an extension of it or them in such a way as to allow no role for imagination in ordinary routine perception.[27] But we should consider how it would be possible to give a kind of caricature-explanation on different lines. Of course, the explanation would run, this indispensable faculty of imagination is involved in ordinary routine perception. It is just that it would be highly misleading to single it out for mention as responsible for the outcome in the case of ordinary routine perception. For to do so would be to suggest that things are not as they normally are in ordinary routine perception. Thus we single the faculty out for mention when it operates without anything like the normal sensory stimulus altogether, as in imaging, delivering mental products unmistakably different from those of ordinary perceptions; when, in one or another of many possible ways, it deviates from, or adds to, the response which we have come to consider routine; or when, as in the present case, we actually mistake the character of the source of stimulus. But it is absurd to conclude that because we only *name* the faculty in these cases, the faculty we then name is only operative in these cases. We might as well say that the faculty of verbalizing or uttering words is not exercised in intelligent conversation on the ground that we generally say things like 'He was verbalizing freely' or 'He uttered a lot of words' only when, for example, we mean that there was no sense or point in what he said.

It is not my purpose to represent such a line of argument as correct.[28] Still less am I concerned – even if I could do so – to elaborate or defend any account of what we really mean, or ought to mean, by 'imagination', such as that line of argument might point to. I am not sure that either the question, what we *really* do *mean* by the word, or the question, what we *ought* to mean by it, are quite the right ones to ask in this particular case. What matters is that we should have a just sense of the very various and subtle connections, continuities and affinities, as well as differences, which exist in this area. The affinities between the image-having power and the power of ordinary perceptual recognition; the continuities between inventive or extended or playful concept-application and ordinary concept-application in perception: these are some things of which we may have a juster sense as a result of reflection on Kant's use of the term 'imagination'; even, in the latter case, as a result of reflection upon Hume's use of the term. A perspicuous and thorough survey of the area is, as far as I know, something that does not exist; though Wittgenstein's pages contain an intentionally unsystematic assemblage of some materials for such a survey.

[1] Immanuel Kant, *Critique of Pure Reason*, trans. Norman Kemp Smith (London: Macmillan, 1929), A 120n.

[2] Kant, B 278–9 (my italics).

[3] Kant, A 141/B 180–1.

[4] David Hume, *A Treatise of Human Nature*, ed. L. A. Selby-Bigge (Oxford: Clarendon Press, 1888), p. 24. Spelling and punctuation have been changed.

[5] Hume, p. 189.

[6] Hume, p. 212.

[7] I modify at least the appearance of Hume's argument here. He seems to suppose that the required premise at this point has an *empirical* character.

[8] Kant, A 120.

[9] Perhaps it is necessary to add that I do not mean that we could not conceive of any circumstances at all (for example, of mental disorder) in which this would be an apt thing to say.

[10] Ludwig Wittgenstein, *Philosophical Investigations*, trans. G. E. M. Anscombe (Oxford: Blackwell, 1953), pp. 204, 212, 213–14.

[11] Wittgenstein, pp. 193–4.

[12] Wittgenstein, p. 197.

[13] Ibid.

[14] Wittgenstein, p. 212.

[15] Wittgenstein resists the generalization. See p. 197. ' "Seeing as" is not part of perception', and p. 195. But he also gives part of the reason for making it; see pp. 194–5.

[16] Wittgenstein, p. 197.

[17] Cf. p. 194; and p. 207: 'Those two aspects of the double cross might be reported simply by pointing alternately to an isolated white and an isolated black cross.'

[18] Wittgenstein, p. 197. (my italics).

[19] Wittgenstein, p. 212.

[20] Wittgenstein, p. 200.

[21] Wittgenstein, p. 207.

[22] Ibid.

[23] Wittgenstein, p. 213.

[24] Ibid.

[25] See p. 53 of Miss Ishiguro's admirable treatment of the whole subject in 'Imagination', *Proceedings of the Aristotelian Society*, suppl. vol. XLI, (1967).

[26] See Miss Anscombe's 'The Intentionality of Sensation: A Grammatical Feature' in *Analytical Philosophy*, 2nd ser., ed. R. J. Butler (Oxford: Blackwell, 1965), 155–80.

[27] As, for example, by saying that when what presents itself as a perception (or memory) turns out to be erroneous, we *reclassify* it by assigning it to that faculty of which the essential role is, say, unfettered invention; somewhat as we sometimes refer to falsehood as *fiction*.

[28] It would be, it will be seen, an application (or misapplication) of a principle due to H. P. Grice. See 'The Causal Theory of Perception', *Proceedings of the Aristotelian Society*, suppl. vol. XXXV (1961), 121–68.

KANT'S CATEGORIES AND
THEIR SCHEMATISM

LAUCHLAN CHIPMAN

IN most editions of Kant's *Critique of Pure Reason* the section entitled 'The Schematism of the Pure Concepts of Understanding' takes no more than eight pages. It is beyond dispute that Kant regarded this short section as constituting an essential part of the proof that pure concepts of understanding or categories play a necessary role in the production of any experience we can conceive ourselves as having. Whether Kant was right to so regard it has been disputed. I shall argue that he was. In the first part of the paper I shall offer an exposition of the Transcendental Deduction, in an attempt to show that a number of questions which arise out of it, and which Kant was entitled to regard as important, are left unanswered at the conclusion of the Analytic of Concepts. The Schematism is intended to answer one such question.

In the second part I shall set out and examine what I take to be the primary doctrines of the Schematism section, and shall argue that although far from defensible in every respect, the Schematism at least has the merit of being a largely coherent treatment of a problem which is real, and which can be raised quite legitimately even apart from the context provided by the *Critique*. In the course of setting out my interpretation I shall argue that the Schematism has been significantly misunderstood by a number of recent and contemporary commentators.[1]

I THE NEED FOR THE SCHEMATISM

Kant introduces the Schematism as part of the Analytic of Principles, it being assumed that the Analytic of Concepts has already demonstrated (a) that there are categories and what categories there are, and (b) that the categories must necessarily be invoked by the imagination in its task of converting a sensory given into a condition such that (A) its owner can think

From *Kant-Studien* 63 (1972), pp. 36–50. Reproduced by permission of the author.

of what appears to him as *his,* and (B) its owner can think of *what appears to him* as his.

In effecting this condition under the description (A), the imagination, according to Kant, has created – for a given subject of sensory representations – the empirical possibility of awareness of the analytic truth that 'all my experiences are mine'. This possibility is brought about by the fact that the condition in question is the very same condition as is required for an interpretation of the token-reflexive expression 'my'. For in order to 'think the analytic unity of apperception', i.e., to entertain the analytic proposition that all my experiences are mine, I have to perform a unique referential or identificatory task, enjoined by the use of the first person expression.

The possibility of any such first-person use has its grounds in whatever grounds the possibility of a use which is not first-personal, and it is this which is emphasized under the description (B). For to think of there being something which appears to me – a necessary condition for my being able to think of myself – I must think of what appears to me as not in principle exhausted by its appearing to me. Were I to think of it as thereby in principle exhausted, I could not think of it as *something* or, in Kant's preferred phrase, 'an object in general', at all. For the conceptual minimum that is involved in the notion of an object is the idea of a 'collector' of characteristics which, very simply, is the idea that at any given time there at least *might* be more to what is given than what, at that time, 'meets the eye'. This minimal conception of an object is sufficiently minimal to be consistent with its *esse* being *percipi*, for what does not 'meet the eye' now might be thought of as nothing but what would meet the eye, given some condition.

Thus it is that, according to Kant, the conception of myself and the conception of an object, and hence of an objective world, enter together. I cannot think of something as given to me in experience without being able to think of the experience in question as *mine*, but to think of it as mine I must *ipso facto* think of what is given as objective; i.e., as belonging to or constituting a field in which the above minimal notion of an object has application.

The concepts necessary for objectivity must accordingly derive from the understanding exclusively. They could not be derived from experience for their possession is its necessary prerequisite. Experience thus requires, according to Kant, the employment of pure concepts of the understanding, or categories. The utilization of the categories in combining or synthesising the manifold of representations is a task with which the imagination is charged. The manifold of representations must, on its part, possess an *affinity* or capacity for entering into just such combinations or syntheses as imagination is required to forge. In short the imagination, operating under the aegis of the understanding, combines or synthesises the manifold of representations into objective appearance in accordance with the understanding's stock of concepts not borrowed from experience; i.e., the pure concepts or categories.

The soundness or unsoundness of the above argument does not concern us here and now. The point of this cursory and sympathetically selective summary of the Transcendental Deduction of the categories is to enable us to determine precisely what has and what has not been established, *if* the argument is sound. By being clear as to what Kant is entitled to regard as settled before the Schematism is introduced, we are better placed to determine what questions the Schematism is designed to answer.

If sound, the Transcendental Deduction establishes that it is a logically necessary condition of my having the concept of myself as a subject of experiences – and hence, it might be argued, of my having the concept of experience at all – that I can apply and have applied to the matter of sense at least one concept which was not extracted from it. What it manifestly does not establish is the number and character of such pure concepts or categories. There is another argument which is supposed to give us just this information, and that of course is the so-called Metaphysical Deduction. It lists what is reputed to be an exhaustive enumeration of all the categories, and provides us with a hint of insight into their distinct characters in terms of our antecedent understanding of the distinct logical relations from which their names are inherited.

Given that the Metaphysical Deduction is also sound we are entitled to conclude that experience is possible only if the understanding possesses at least one of the enumerated categories and, in its capacity as imagination, accordingly synthesises the manifold of representation. This is the optimum combined yield of the two Deductions. They do not prove that *all* the categories are necessary for the possibility of experience, although Kant often speaks loosely as if he believed they do. It is formally consistent with the story so far that only one of the enumerated categories is ever in fact employed. Indeed it has not yet been established, concerning any *particular* category, that *that* category is necessary for the possibility of experience nor even, concerning any particular category, that that category *can* guide the imagination in the synthesis of representations. Kant does cite the categories of quantity and cause (in B 162 and B 163) to illustrate his conclusion that all synthesis is subject to the categories, but in these passages it is simply *asserted* that synthesis conforms with them; there is, in this context, no proof of their necessity nor demonstration of their possibility.

Kant's ultimate aim is to prove the necessary involvement of *all* of the enumerated categories in experience. By noticing how far short of a full proof of this thesis we are at the conclusion of the Analytic of Concepts, we place ourselves in a sympathetic position for understanding the project of the Schematism. The question may be put simply thus: what do we need to add to the Metaphysical and Transcendental Deductions, assuming their soundness, to reach Kant's thesis? The answer is a proof, for each category, of its necessary involvement in experience. And if there is some general objection to the implied claim that a category can be involved in experience in the required way, we should expect it to be cleared away either as a prelude to, or in the course of, any such proof.

This is exactly what we find. Prior to a discussion of each of the categories in turn, Kant devotes a section to trying to prove that all categories can operate with imagination in the way that the Transcendental Deduction requires, and enter into an 'objectifying' of sensory data. To do this, Kant attempts to show that a necessary condition which, he believes, must be fulfilled for any application of any *empirical* concept to the data of inner or outer sense, is also fulfilled, despite a prima-facie case to the contrary, in the application of any pure concept. He then proceeds to show how this requirement is fulfilled for each of the twelve categories. All of this constitutes a prelude to the proofs, for each category in turn, of their necessary involvement in experience. The prelude is the Schematism.

The necessary condition which, Kant believes, must be fulfilled is that a concept can have application only insofar as there is a *homogeneity* between the concept applied and the data to which it is applied: 'in other words, the concept must contain something which is represented in the object that is to be subsumed under it'.[2] In the case of empirical concepts, this seems to mean that it must be possible to provide an elucidation of any such concept which would issue in a list of simple descriptive predicates, perhaps embodied in statements of necessary and/or sufficient conditions for something's falling under the concept, which characterise any sensory data to which the concept has application. We might think of these descriptive predicates as representing the *elementary sensory components* of a concept. Kant's view seems to be that it is possible to apply empirical concepts to their instances because such concepts possess elementary sensory components which correspond to sensible features of the data which fall under the concepts.

Empirical concepts and their instances are thus homogeneous. Kant's examples – the concepts of a plate, a dog, and a triangle – are all of middling complexity. No doubt he would hold that the same requirement of homogeneity must be met by highly sophisticated and complex concepts such as those of a government or an institution, and by very simple concepts, such as the concept of red, whose only component is the sensory component represented by the predicate 'red'.

How far the above observations are true, let alone useful as an account or part of an account of what is involved in the application of an empirical concept to its instances, is a matter which cannot be gone into here. They are offered only as an elucidation of Kant's obscure talk of homogeneity and not as a defence of it.

Pure concepts do not contain elementary components. There is nothing something must look like if it is to look like a cause or a possibility. The application of pure concepts, it seems, must be in virtue of something other than a matching between the data falling under them and their constituent sensory components, for they have no sensory components. It should be noticed that this does not mean that we can never look at, point to, or 'intuit' – in Kant's sense – instances of some pure concept. For clearly we can and do point to causes, unities, and actualities, for example.

G. J. Warnock, whose interpretation of Kant is so far in several respects similar to mine, seems to be confused on just this matter. Concerning the heterogeneity of the pure concepts of unity, causality, and possibility, Warnock writes:

Clearly, 'This is one' (if allowed at all) is not like 'This is white'; 'This is the cause' is not like 'This is the football'; a possible President does not, at the moment of electoral triumph, lose one characteristic, possibility, and acquire a new one, actuality. What is referred to by 'one', or 'cause', or 'possible', is in no case a thing that I can look at, point to, 'intuit'; I cannot, then, learn to use these words in the same way as I learn to use 'white' or 'round'.[3]

Implicit in Warnock's last sentence seems to be the suggestion that instances of pure concepts can never be perceived, but such a suggestion is surely mistaken. For while it is perhaps not possible to construct comparable locutions for all of the categories, it is certainly possible to use perceptual verbs directly in connection with some of them. For example, having seen the flash of lightning which caused the bridge to be burnt down, one could correctly say one saw the cause of the fire which destroyed the bridge.

Warnock is nonetheless right in thinking that for Kant, the heterogeneity of pure concepts and sensory data is connected with perception. It is not that one cannot intuit causes, substances, pluralities, and actualities, for example. Rather, it is that although, like empirical concepts, some pure concepts can be applied directly to the representations of sense, pure concepts are never applied solely on the basis of the sensory given. One can call something a dog because of what it looks like – it presents a doggish appearance – but one cannot call something a cause because it presents a cause-ish appearance! A simple corollary of the proposition that pure concepts do not contain sensory components is the impossibility of applying a pure concept *solely on the basis of* the sensory given. None of this entails that a pure concept cannot be applied *to* the sensory given.

The confusion with which I have charged Warnock occurs at one point in Kant's own writings. In stating the problem with which Schematism is intended to deal Kant writes:

But pure concepts of understanding being quite heterogeneous from empirical intuitions, and indeed from all sensible intuitions, can never be met with in any intuition. For no one will say that a category, such as that of causality, can be intuited through sense and is itself contained in appearance. How, then, is the *subsumption* of intuitions under pure concepts, the *application* of a category to appearances, possible?[4]

Taken literally, this passage is simply self-contradictory. The presupposition of the question with which it concludes is that pure concepts do apply to intuitions, which surely means that they can be instantiated in appearances. Yet this is precisely what the first quoted sentence would seem to be denying.

The passage is not to be taken literally, but as an awkward expression of what is most clearly expressed in the second half of the second quoted sentence. What Kant is most emphatically denying is that a pure concept might apply to appearances in virtue of the phenomenal character of those appearances. Were this so, we could think of the question of whether an application of a pure concept on a given occasion is justified as something which could at least in large part be resolved by inspecting the subsumed appearances, much as one can tell whether someone was justified in calling something a dog by having a look at what he called a dog. Kant is most coherently construed as denying that whatever *grounds* or *justifies* the application of a category to appearance 'can be intuited through sense and is itself contained in sense'; i.e., in terms with which we are now familiar, as denying that one can ever justify the application of a pure concept to appearance by showing that some set of sensory components, contained in the concept, is satisfied by the appearance in question.

Hence Kant's problem. If pure concepts are not homogeneous with the data of sense to which they apply, then this seems as good a reason as any for saying that they have no real part in experience at all; that they are, perhaps, illegitimate postulations or anthropomorphic illusions extravagantly mis-referred to appearances. If pure concepts do not contain sensory components, then this seems as good a reason as any for saying that they are not homogeneous with the data of sense. Kant thus finds himself confronted with a very strong prima-facie case against the thesis that the categories are far more than 'empty logical forms' enjoying analytic relations with one another. The Schematism is designed to meet this prima-facie case. Unless it can be demonstrated that concepts not containing sensory components

can be utilized by the imagination in its work of synthesis, the argumentation of the Analytic of Concepts is entirely in vain. We would be left with
nothing but a *reductio ad absurdum* of the Transcendental Deduction.

The Schematism is thus intended to answer one of the questions left
unanswered by the Analytic of Concepts; namely, how is the subsumption
of appearances under pure concepts possible? I have attempted to show
why Kant is right to regard this section as important. Its importance in the
Critique is both prospective and retrospective. Were the question of the
Schematism not to be answered satisfactorily, the subsequent attempts to
prove, for each category in turn, the necessity of its involvement in experience, would be idle, for the possibility on which they each depend would
then look to be unfulfilled. Equally, the Transcendental Deduction itself
would be undermined, for the conclusion to which it leads would then look
to be, in retrospect, an impossible one.

II THE DOCTRINES OF THE SCHEMATISM

Kant proposes to solve the problem of showing that it is possible for each
pure concept to apply to or subsume under itself intuitions by showing *how*
this is possible; i.e., by producing a theory which has the consequence that
the heterogeneity of pure concepts and appearances does not preclude the
former from applying to the latter. Kant's 'solution' is to hold that, while not
homogeneous with each other, there is nonetheless something with which
both category and sensory data are homogeneous, and this Kant calls the
transcendental schema. Construing homogeneity as non-transitive, Kant is
able to hold that category and sensory data are connected by a link with
which they are alike homogeneous, without having the consequence that
category and sensory data are homogeneous with each other. Category and
transcendental schema are alike *a priori* in origin but the schema, because of
the role to which Kant assigns it in the process of synthesising sensory data,
is also appropriately called sensible. Nothing in the shared purity of the
category and the schema corresponds to anything in the sensory data nor,
according to Kant, is the shared sensibility of the sensory data and the
schema reflected in anything in the category.

What does all this mean? The notion of a schema is probably most easily
grasped in connection with empirical concepts. The notion of a transcendental schema, or schema appropriate to pure concepts, can then be
understood derivatively. Kant says very little about the schemata of empirical concepts; so little that it is tempting to conjecture that the doctrine was
developed primarily with pure concepts in mind, and then half-heartedly

generalised in order not to look excessively *ad hoc*. Nonetheless what he does say, meagre and equivocal as it is, together with what he says about transcendental schemata, is suggestive of a comprehensive general theory of schemata, and it is this which I shall attempt to elaborate.

A schema is a rule or set of rules which, in the case of the empirical concepts, are acquired *a posteriori*. Such rules are employed by the imagination in combining or synthesising sensory data. If I have the empirical concept of a dog, and can apply it to its instances, then, according to Kant, I must be able to combine or synthesise my manifold in any of the indefinitely large number of different ways necessary to enable me to identify any dog with which I am presented as a dog, or to imagine any dog described to me in the absence of the dog described.

Empirical concepts are, for Kant, unique in being identical with their schemata. Kant writes:

The concept 'dog' signifies a rule according to which my imagination can delineate the figure of a four-footed animal in a general manner, without limitation to any single determinate figure such as experience, or any possible image that I can represent *in concreto*, actually presents.[5]

Given that a schema is a 'representation of a universal procedure of imagination in providing an image for a concept', it is clear from the above that the concept of a dog, and its schema, are one and the same.[6]

This is not the case with *pure* sensible concepts, nor with categories. In the case of a pure sensible concept, such as the concept of a triangle, the schema and the concept cannot be identical, since the concept is not of something that can be met with in experience. More precisely, although we do encounter things that are more or less triangular, we do not encounter things which are triangles. Given that the concept of a triangle cannot itself be instantiated in our sensory data, in contrast to an empirical concept such as that of a dog, how can we legitimately describe what does confront us as triangular in shape, or even roughly triangular? We can do this, according to Kant, because associated with the concept of a triangle is a rule for synthesising sensory data. One who is in possession of this rule, which is 'a rule of synthesis of the imagination, in respect to pure figures in space', is able to synthesise the manifold of sensory data in any of the indefinitely large number of ways necessary to enable one to identify any triangular shape one encounters as triangular.[7] This is the schema of the concept of a triangle.

Unfortunately, Kant nowhere makes known to us what the nature of this association is. There is no clear sense in which the schema is *derivable* from the concept, for there is no reason to suppose that the rule which is the schema is even statable. We are still left with the problem, how is it possible to synthesise one's manifold in accordance with a pure sensible concept, such as a geometrical concept? All we have been told is that it is because we are in possession of a rule which enables us to do just that! The problem is not a new one, but simply Plato's problem of giving an account of how it is possible to subsume a concrete particular under a thoroughly abstract universal. Kant's solution involves locating the abstract universal and the mechanism of subsumption in us, and viewing the mechanism of subsumption as essentially 'constructive' in character. What this solution does not tell us is the nature of this mechanism of subsumption and, more importantly, its relation to its parent universal. Of course this is not to say that no adequate account can be given.

Before proceeding to Kant's account of schemata in connection with the categories, let us review the theory as it has emerged so far. I have given one reason for finding it unsatisfactory in connection with pure sensible concepts. I propose now to argue that it is unsatisfactory even for empirical concepts; in particular, that Kant should not have identified empirical concepts with their schemata. I shall also argue that although far from satisfactory, Kant's theory is nevertheless not vulnerable to one of the more important and now standard objections brought against it, to the effect that Kant's is simply a sophisticated variant of a resemblance or replica theory of concept instantiation. My conclusion will be that Kant's general theory of schemata, inadequate as it is, nonetheless can be construed, with relatively minor modifications, as a coherent attempt to deal with a real problem. Whether it can be adapted to deal with Kant's problem about the subsumption of sensory data under the categories is another matter.

If empirical concepts were identical with their schemata; if, for example, the concept of a dog were identical with its schema, then the following propositions would constitute an inconsistent triad:

(i) Jones possesses the concept of a dog
(ii) Jones is currently attending to a sensory field in which is contained a paradigmatic instance of the concept of a dog
(iii) Jones cannot tell that what is before him is a dog.

On the other hand, if these three propositions are co-tenable, the concept of a dog is not its own schema.

Jonathan Bennett, it seems, would maintain that they do form an inconsistent triad and hence, if there is any point in talking of schemata for empirical concepts at all, they had better be identified with those concepts.[8] Bennett explicitly rejects Warnock's charge that Kant illegitimately separated the application of concepts from having them, on the ground that if a person's inability to apply a concept were due to (a) the fact that it had no instances, or (b) the fact that a sensory disability prevented him from recognizing its instances, this would not entail that he did not possess the concept. Possession of a concept is, for Bennett, sometimes consistent with the inability to apply it; it depends on the source of the inability.

If a concept does have instances, and the person who claims to possess it does not have a relevant sensory disability, then, according to Bennett, he must be able to apply it. If (iii) in our triad is false, then so, too, is at least one of (i) and (ii). Bennett's only argument for this is fallacious. Bennett writes:

> But I could not possess a concept yet be unable to apply it because of an intellectual defect, a defect in my 'judgement' which is 'one of the higher faculties of knowledge'. Having a concept involves being able both to use it in 'rules' and, under favourable sensory circumstances, to apply it to its instances. You will not credit me with having the concept of a dog just because I can state many general truths about dogs, such as that they are mammals, never laugh, have legs, etc. If I can do this and yet – although not sensorily disabled – apply the word 'dog' to particular birds, humans, porpoises, etc., and often apply 'not a dog' to particular dogs, you must conclude that I do not understand 'mammals', 'laugh', 'legs' etc. But in that case my stock of 'general truths about dogs' is like a parrot's repertoire: it is not evidence that I understand the word 'dog' in any way at all.[9]

It is true that the fact that a person can utter a number of general truths embodying a given concept does not, in general, guarantee that he possesses that concept. It is also true that the fact that a person regularly *misapplies* a given concept does, in general, guarantee that he does *not* possess that concept. But the fact that we can legitimately infer a person's non-possession of a concept from his regular misapplication of it, does not in the least imply that we can legitimately draw the same conclusion from his regular *non*-application of it, even when this non-application cannot be put down to the absence of suitable instances or to a sensory disability. Yet this is precisely what Bennett needs to establish his thesis that 'having a concept involves being able . . . under favourable sensory circumstances, to apply it to its instances'. Bennett's argument thus cannot be used to show that our triad is inconsistent.

Any appearance of inconsistency which the triad possesses is actually a consequence of the fact that ours is a dog infested culture, as I shall now try to show. Substitute for 'a dog' in each of (i), (ii) and (iii), either the expression 'a tadpole' or the expression 'bone marrow', and the triad is plainly not inconsistent. If a person knew that the tadpole stage was part of the life cycle of a frog, knew which part of the life cycle, knew something of how tadpoles are produced, where they are to be found, and that they are very small, then we would correctly say that he had the concept of a tadpole. Of course, we would have to revise our assessment if he then went on to call goldfish or cockroaches tadpoles, or denied tadpolehood of its paradigm instances. Regular or glaring misapplication is not allowed. But if, confronted with a jar of water containing tadpoles, he confessed his ignorance as to what those things in the water are, we would be under no obligation to revise our assessment. His ignorance would not have to be described as conceptual ignorance. He knows what tadpoles are, but cannot recognise them when he sees them.

The example of bone marrow makes the same point over again, perhaps more persuasively. Most people who have the concept of bone marrow would not recognise it if confronted with it. It seems that we can distinguish, in Kant's language, the concept of a tadpole or the concept of bone marrow from the relevant schemata. The capacity to synthesise their manifolds of sensory data in accordance with these concepts does not belong to all those who have the concepts. This is a ground for distinguishing the concept and the schema, for it is certainly possible – in fact often the case that one should have the former without having the latter.

Why is the example of the concept of a dog so persuasive the other way? Simply, I think, because dogs are unavoidable in ways that tadpoles and bone marrow are not. Our situation is one in which we are frequently required to discriminately respond to the presence of dogs. The concept of a dog is one that is heavily worked in quite close proximity to areas of sensory discrimination; we all, from time to time, find ourselves in the position of having to look for, look at, buy, bathe, give away, or run away from, dogs. It is difficult to imagine someone who possessed the concept of a dog and suffered no sensory disability not having *had* to apply it on some occasion or other.

Difficult, but not impossible. We could imagine a person whose background experience contained amazingly few or no actual encounters with dogs, but a good deal of 'dog theory'. The case for saying of such a person that he has the concept of a dog is at least as good as that for saying of some of us that we have the concept of a tadpole, and of most of us that we have the concept of bone marrow. If the case is allowed, our triad is not inconsistent. If our triad is consistent, Kant was wrong to identify empirical concepts with their schemata. Although empirical concepts differ from pure sensible concepts in that instances of the latter are not to be met with among the sensory data, Kant should have treated them similarly from the point of view of their schemata. In neither case should the concept be construed as a rule for combining or synthesising the sensory data in certain ways, for it is in no way necessary that one who is in possession of a concept is in possession of some such recognitional rule.

A consequence of divorcing empirical concepts from their schemata in the way that I have suggested is that the same problem as arose concerning the relation between pure sensible concepts and their schemata now arises for all empirical concepts. For what are we to say is the relation between the schema for the concept of a dog, and the concept, if it is not identity? This is a real problem, but it is not peculiar to Kant. It is, as has already been noted, the classical problem of giving an adequate account of the subsumption of a particular under a universal.

Kant is commonly held to have given a sophisticated version of a classically wrong solution to this classical problem. The unsophisticated version is that universals are concepts, concepts are private mental images, and subsumption is the favourable outcome of an exercise in comparing an object with a private mental image. Kant's is sophisticated, it is said, in that he does not think of concepts as *being* mental images, nor does he think of there being *just one* mental image associated with each concept. Rather, Kant thinks of each concept as either identical with or associated with just one schema, and of schemata as rules for the generation of private mental images.

The passage which is usually quoted in support of this interpretation and which, superficially read, seems to support it conclusively, is one in which Kant appears to be defining a schema. Kant writes:

The representation of a universal procedure of imagination in providing an image for a concept, I entitle the schema of this concept.[10]

Both Warnock and Bennett interpret this passage as indicating that Kant regarded schemata, at least in the case of empirical concepts, as rules for constructing private images to serve as objects of comparison in making applications of concepts. Bennett writes:

Kant wants his schematism theory, I think, to explain how we are able to recognize, classify, describe. For example: I have no doubt that this thing here in front of me is a dog; but what, for me now, links *this* with other things which I have called 'dogs', in such a way that I am entitled to call this a dog too? Kant's answer is that I can link this dog with other dogs by conjuring up a mental picture of a dog checking it against the object which I now see. I know that my mental picture is of a dog because I have produced it in accordance with the schema of the concept of a dog. If the thing in front of me is indeed a dog, then an adequate schema of the concept of a dog will generate – as well as many images which do not help me with my present problem – at least one image which corresponds closely enough to the object now before me to justify my going ahead and calling the object a dog.[11]

Bennett then, correctly, objects to this as an account of how we are enabled to apply the concept of a dog, using familiar arguments. Warnock offers similar objections.

Both Warnock and Bennett make the same mistake of thinking that Kant means by 'image' in this context much the same as is meant by the same word in contemporary ordinary usage. He does not. The word as used by Kant functions as a noun for anything which is brought about as a result of the operations of the faculty of imagination. It therefore comprehends, but is in no way restricted to, private mental pictures. Whether synthesis or combination of the sensory data occurs, Kant calls the result an image. This somewhat technical use of 'image' is plain from the first edition Transcendental Deduction, where in a passage explaining the role of the imagination in synthesis, Kant writes:

There must therefore exist in us an active faculty for the synthesis of this manifold. To this faculty I give the title, imagination. Its action, when immediately directed upon perceptions, I entitle apprehension. Since imagination has to bring the manifold of intuition into the form of an image, it must previously have taken the impressions up into its activity, that is, have apprehended them.[12]

There is accordingly no reason for reading Kant as holding that schemata, being rules for synthesis employed by the imagination, produce images *only* in the contemporary ordinary sense. It is consequently inappropriate to bring against Kant's account the standard objections to replica or resemblance theories of concept application, for his is not such an account, sophisticated or otherwise.

So far we have seen no reason to regard Kant's doctrine of schemata as any less coherent than the notion of synthesis, on which it is evidently parasitic. The latter notion is notoriously difficult because of the problems of giving any elucidation of it which does not reduce to a set of spatial and chemical analogies, but that is another problem. That Kant's general doctrine of schemata is relatively coherent does not, of course, entail that his extensions of it to handle pure concepts are also coherent, and it is to these we now turn.

In essence, the extensions involve claiming that associated with each category there is a transcendental schema, which is a transcendental determination of time. Kant offers only one argument as giving sense and support to his claim.[13] I shall expound what I take the argument to be. To follow the argument, we need to make two concessions to Kant. First, let us concede that Kant has shown that time is the form of all inner sense or, as he expresses it here, 'the formal condition of the manifold of inner sense'. Kant takes this to license him in speaking of time as *containing* an *a priori* manifold in intuition. What Kant means by this is quite unclear, but whatever it means he takes it to imply that *a priori* general propositions about time are, if true, true of any possible inner or outer sensory field, and therefore have to be acknowledged in any synthesis of a sensory field, no matter what *a posteriori* schematic dispositions of its constituent data happen to be appropriate on a given occasion.

Second, Kant requires us to concede that a concept of the understanding 'contains pure synthetic unity of the manifold in general'.[14] However, the only reason we could have for conceding this is the argument of the

Transcendental Deduction, in so far as it is supposed to show that some-thing non-empirical must work on the sensory manifold before it becomes experiential; i.e., before it becomes such that a person could conceive of *himself* in an objective world. At this point, the question which it is the prime purpose of the Schematism to answer looks to be crucially begged. We are not entitled to use this conclusion of the Transcendental Deduction, it could be argued, for precisely what is in issue is whether a category *can* be the non-empirical something which the Transcendental Deduction repu-tedly shows to be *a priori* necessary.

The argument of the Schematism thus makes use of the idea that a category brings it about that the manifold can be thought of as all experienced by me, as objective, and as permitting self-awareness ('con-tains pure synthetic unity of the manifold in general'). Hence it might be objected that, rather than grounding the legitimacy of the employment of the categories to 'make experience possible', it presupposes it.

The objection rests on a misunderstanding of Kant's methodology. As we have seen, Kant's object in the Schematism is to show that the subsumption of appearances under categories is possible, and the method chosen is that of showing how this is possible. In attempting to show *how* something is possible it is perfectly legitimate to conditionally assume that it is in fact the case, and then go on to show how what has been assumed to be the case can be made to integrate in a theoretically acceptable way with less contentious doctrines. When this is done, the possibility of what was assumed has been demonstrated, the demonstration having shown how what was assumed to be the case could be the case – it admits of integration with accepted or acceptable theory.

Having made these concessions to Kant, we must ask why Kant says that transcendental schemata are transcendental determinations of time, varying from category to category. Transcendental determinations of time are chosen as transcendental schemata because they have something in common with the categories, and with appearances, without being identical with either. This is supposed to come about in the following way. From the Aesthetic, we know that no appearance can occur other than in time. Hence any general truth to do with time will manifest itself in any appearance; e.g., through a necessary rule of synthesis. Categories are necessary to pass from raw data to appearances, and categories are agreed to be *a priori* in origin. Accordingly, since any synthesis requires a category (authority: Transcendental Deduction), and since any synthesis will necessarily yield a temporally ordered product (authority: Aesthetic), the schema appropriate to each category may be thought of in the following way. Each is a rule for ensuring that any organisation of the sensory data will make it necessary for the subject to think of *whatever* product emerges as dependent for its objectivity on its temporal properties.

The necessity referred to is that of employing temporal conceptions in any thought of appearance as more than subjective representations. Transcendental schemata are not supposed to be rules for synthesising the manifold of representations into appearances of *particular* objects, for it is the schemata of empirical concepts which have this role. Rather, transcendental schemata are rules to ensure that the general and invariable conditions of objectivity, all of which are explicable in ways which involve essential reference to time, are fulfilled in any particular synthesis. In its weakest and perhaps most defensible form, the thesis is that transcendental schemata are rules which require that certain temporal questions be answerable in consequence of any synthetic 'objectifying'.[15]

What has been offered above is, at best, a picture of the intelligibility of the doctrine of transcendental schemata, rather than a demonstration that the doctrine is intelligible. The real test comes with the working out of the schema for each of the categories. Unfortunately, it is only in respect of the relational categories of substance, causality, and interaction, that the formulations of the transcendental schemata seem other than absolutely *ad hoc*. The schema of substance is said to be permanence to the real in time. True, to think of something as substantial necessarily involves thinking of it as possessed of some sort of history, which means that one must think of it as extending through some period of time such that, at any time during that period, whether or not it manifested itself among my sensory data, it nonetheless existed at that time. Instantaneous substance seems to be out of the question. Causality and interaction relate to time in ways developed in the Second and Third Analogies, and to consider their schemata would involve trespassing into those areas.

With the categories of modality, quantity, and quality, the schemata become progressively less plausible. For example, the pure schema of quantity or magnitude is supposed to be number – a consequence of the fact that any quantity is measurable and therefore contains countable units, and counting, construed as enumeration, takes time.[16]

If there is any truth at all in the doctrine of transcendental schemata, it seems to be only that applications of some of the categories carry temporal commitments, and hence if Kant is right in maintaining that the categories are antecedent to and necessary for the possibility of experience, then not only must we conceive of experience as temporally ordered, but we must also 'objectify' the data of experience, whatever the data may happen to be, and whatever empirical concepts we may impose, in such a way as to invoke this necessary orderliness.

This is an appealing doctrine, but it cannot be said that Kant has done a great deal in the Schematism to furnish it with support. To point this out, however, is not to bring a legitimate criticism against the Schematism in particular, for the task of *proving* the necessary involvement of each of the transcendental schemata in synthesis belongs to the Axioms, Anticipations, Analogies, and Postulates. The task of the Schematism was simply to show how the categories *can* be applied to appearances. At best, what has been shown is that by construing each category as having a temporal dimension, and assuming the necessary temporality of outer and inner sense has already been established, we guarantee its applicability to sense. Its necessary applicability is another matter. Where the argument of the Schematism is least convincing, sometimes because there is simply no argument at all, is in establishing, for each category, its temporal construction.

In this section I have tried to give a sympathetic elucidation of the doctrines of the Schematism. I have argued that there is a general problem about the subsumption of appearances under concepts which is by no means limited, as Kant seems to have thought, to pure concepts. Although Kant's theory, that we are enabled to subsume appearances under pure concepts because we are in possession of rules for doing just that, says very little, it is coherent – provided that we do not, unreasonably, require that the rules be statable. Surprising at it may seem, under certain interpretations, the theory is empirically testable. For like many of Kant's doctrines, it admits of an unintended physicalistic interpretation, according to which the schemata could be construed as neural mechanisms for bringing about interpretations of sensory input, with concepts construed as relatively permanent structural conditions of the thinking centres. Of course this is sheer speculation, but then Kant wrote of the Schematism that:

. . . in its application to appearances and their mere form, [it] is an art concealed in the depths of the human soul, whose real modes of activity nature is hardly ever likely to allow us to discover, and to have open to our gaze.[17]

[1] In particular, by G. J. Warnock, 'Concepts and Schematism,' *Analysis* IX (1948–9), 77 ff. and J. Bennett, *Kant's Analytic* (Cambridge University Press, 1966), ch. 10.
[2] Kant, *Critique of Pure Reason*, A 137/B 176.
[3] Warnock, 'Concepts and Schematism', 81.
[4] Kant, A 138/B 177.
[5] Kant, A 141/B 180.
[6] Kant, A 140/B 180. Provided that the word 'image' is not misinterpreted to mean a private intermediary between the sensory data and ourselves. This point is taken up below.
[7] Kant, A 141/B 180.
[8] Bennett, *Kant's Analytic*, 146.
[9] Ibid., 146.
[10] Kant, *Critique of Pure Reason*, A 140/B 180.
[11] Bennett, *Kant's Analytic*, p. 143. See also Warnock, 'Concepts and Schematism', p. 81.

[12] Kant, *Critique of Pure Reason*, A 120.

[13] Ibid., A 138/B 177.

[14] Ibid.

[15] 'We thus find that the schema of each category contains and makes capable of representation only a determination of time.' Ibid., A 145/B 184.

[16] Ibid., A 142–143/B 182.

[17] Ibid., A 141/B 180.

VI

TRANSCENDENTAL ARGUMENTS[1]

BARRY STROUD

IN recent years there has been widespread use of arguments described as Kantian or 'transcendental' which have been thought to be special, and perhaps unique, in various ways. What exactly is a transcendental argument? Before looking closely at some specific candidates it will be useful to see some of the general conditions which such arguments must fulfill.

Kant recognized two distinct questions which can be asked about concepts.[2] The first – the 'question of fact' – amounts to 'How do we come to have this concept, and what is involved in our having it?' This is the task of the 'physiology of the human understanding' as practised by Locke. But even if we knew what experiences or mental operations had been required in order for us to have the concepts we do, Kant's second question – the 'question of right' – would still not have been answered, since we would not yet have established our *right* to, or our *justification* for, the possession and employment of those concepts. Although concepts can be derived from experience by various means, they might still lack 'objective validity', and to show that this is not so is the task of the transcendental deduction.

For example, Kant considered it:

a scandal to philosophy and to human reason in general that the existence of things outside us . . . must be accepted merely on *faith*, and that if anyone thinks good to doubt their existence, we are unable to counter his doubts by any satisfactory proof.[3]

The transcendental deduction (along with the Refutation of Idealism) is supposed to provide just such a proof, and thereby to give a complete answer to the sceptic about the existence of things outside us. We can therefore get some understanding of Kant's question of justification by looking at the challenge presented by the epistemological sceptic.[4]

Since the traditional epistemologist asks how it is possible to know anything at all about the world around us, he is not interested only in the specific question of whether there really is a tomato on the table. Consequently, he will not be answered if we simply appeal to one alleged matter of fact in order to support our claim to know another. You cannot show the

First published in *The Journal of Philosophy* LXV (1968), pp. 241–56; reprinted in Penelhum and MacIntosh (eds.), *The First Critique*. Belmont, Calif.: Wadsworth Publishing Co. Inc., 1969, pp. 54–69. Reprinted by permission of the author and *The Journal of Philosophy*.

sceptic that you're not hallucinating, and hence that you know there is a tomato on the table, simply by asking your wife if she sees it too – hallucinations of your wife's reassuring words are epistemologically no better off than hallucinations of tomatoes. At every point in the attempted justification of a knowledge-claim the sceptic will always have another question yet to be answered, another relevant possibility yet to be dismissed, and so he can't be answered directly.

Doubts about whether some particular hypothesis is true can often be settled by following the ordinary, well-known ways of establishing matters of so-called empirical fact. But the sceptic maintains that the whole structure of practices and beliefs on the basis of which empirical hypotheses are ordinarily 'supported' has not itself been shown to be reliable. As long as we have a public objective world of material objects in space and time to rely on, particular questions about how we know that such-and-such is the case can eventually be settled. But that there is such a world of material objects at all is a matter of contingent fact, and the sceptic challenges us to show how we know it. According to him, any justification for our belief will have to come from within experience, and so no adequate justification can ever be given. Transcendental arguments are supposed to demonstrate the impossibility or illegitimacy of this sceptical challenge by proving that certain concepts are necessary for thought or experience, but before trying to see exactly how they are thought to do this it will be instructive to consider a possible objection to what has been said so far.

If transcendental arguments are meant to answer the sceptic's question, and if, as many believe, that question makes no sense, then there will be little point in considering the exact nature of these alleged arguments. This is reminiscent of the line taken by Carnap.[5] He, like Kant, distinguishes between two types of questions – ordinary empirical questions on the one hand, which are raised and answered from 'within' a framework of concepts, beliefs, and recognized procedures of confirmation, and, on the other hand, questions raised by the sceptic or metaphysician about this framework, raised, so to speak, 'from outside'. To ask whether there are any objects more than ten billion miles from the earth is to ask an 'internal' question to which there is an objectively right answer. It is a genuine 'theoretical' issue which can be settled by discovering the truth of certain empirical statements. But to ask simply whether there are any objects at all is to ask an 'external' question about the existence of the system of spatio-temporal material objects as a whole, and this is not a 'theoretical' question with an objectively right answer at all. It is a 'practical' question, a request for a *decision* as to whether or not we should think and talk in terms of material objects. Since there is no set of true propositions which would answer an 'external' question, the issue cannot be settled by gathering evidence.

The belief that 'external' questions must be answered in the same way as ordinary empirical questions is what leads the epistemologist to the sceptical impasse. Carnap avoids scepticism by denying this and claiming that statements like 'There are material objects' assert nothing about the world at all, and hence that we couldn't conceivably lack knowledge of their truth value. They have no truth value – they merely serve to express a policy we have adopted, or a convention with which we comply.

If this conventionalist line is to be successful there must be no *need* for us to conceive of the world in terms of material objects in space and time; it must be perfectly possible for us to find the world and our experience intelligible in other terms. But transcendental arguments are supposed to prove that certain particular concepts are necessary for experience or thought; they establish the necessity or indispensability of certain concepts. Therefore conventionalism of this sort will be refuted if a sound transcendental argument can be produced. If there are particular concepts which are necessary for thought or experience then it is false that, for every one of our present concepts, we could dispense with it and still find the world or our experience intelligible. A sound transcendental argument therefore would show that it is wrong to think (with the conventionalist) that the only possible justification of our ways of thinking is 'pragmatic' or practical, and equally wrong to think (with the sceptic) that they can be justified only by collecting direct empirical evidence of their reliability. Although these look like difficult demands to meet, they represent the minimum conditions which Kant set for the success of a transcendental argument.

Recent attempts to demonstrate the 'absurd' or 'paradoxical' nature of sceptical questions have taken various forms. It has been argued that seeing a tomato in the clear light of day, when other people say they see it too, when I can reach out and feel it, is simply what we *call* 'finding out there is a tomato there'. This is the best possible case of knowing of the existence of a tomato, and since situations like this certainly do occur, it follows that we *do* know that there are tomatoes, and hence that there are material objects. But from the fact that this is the best possible case of knowing of the existence of a tomato the most that follows is that 'If this isn't a case of knowledge of the external world then nothing is', or, in the more familiar example, 'If this isn't a case of acting of one's own free will then nothing is'. But the truth of such conditionals does not threaten the sceptic; it is precisely because they are true that he is able to challenge all of knowledge by considering only one or two examples. In addition to establishing conditionals of this sort, then, one would also have to show that it's false that there is no knowledge of the external world. But any attempt to show that by an appeal to other empirical facts would lead back onto the sceptic's treadmill.

Defenders of the paradigm case argument failed to see that the sceptic need not deny that we can make all the empirical distinctions that we do make (e.g., between what we call 'hallucinatory' and what we call 'non-hallucinatory' perceptions), or that we all apply certain concepts (e.g., 'of his own free will') in certain circumstances and withhold them in others. In Kant's terms, these are answers to 'questions of fact' and so are not sufficient to answer the 'question of justification'. It is not a sufficient refutation of the sceptic who doubts that p to present him only with a conditional to the effect that if *not-p* we couldn't possibly do A. What is in question is whether we ever 'validly' or 'justifiably' do A. This is shown, in the extreme case, by the obvious weakness of the argument which runs: If no-one ever acted freely, then the ascription of praise and blame would be impossible. But we do ascribe praise and blame. Therefore it is false that no-one ever acts freely.

In order to demonstrate the absurdity of scepticism, the paradigm case argument had to rely on a theory of meaning to the effect that, at least for some words, if those words are to have the meaning they do have in our language, there must actually be things or situations to which they have been, and perhaps still are, truly applied. If this were true of the word 'X', for example, then from the fact that the question 'Are there really any X's?' makes sense it would follow that the answer to it is 'Yes'. This has been thought sufficient to demonstrate the 'absurdity' of the sceptic's question.[6] But this theory of meaning is highly doubtful, for reasons that will be given later. In the meantime I shall examine some subtler and more persuasive recent anti-sceptical arguments.

The first half of Strawson's *Individuals*, which is certainly Kantian in tone, gives the impression of relying on transcendental arguments to establish the absurdity or illegitimacy of various kinds of scepticism. Strawson starts by saying:

(1) We think of the world as containing objective particulars in a single spatio-temporal system.

He emphasizes that this is a remark 'about the way we think of the world, about our conceptual scheme',[7] and he wants to discover some of the necessary conditions of our thinking in this way. In discovering these conditions Strawson claims to have demonstrated that the sceptic's doubts are illegitimate since they amount to a rejection of some of the necessary conditions of the existence of the conceptual scheme within which alone such doubts make sense.[8] This can be understood in two ways, depending on what the sceptic is thought to doubt.

Strawson sometimes takes the sceptic to doubt or deny:

(6) Objects continue to exist unperceived.

Only on this understanding of the sceptic is there any plausibility in the claim that he is merely a 'revisionary' metaphysician who rejects our conceptual scheme and offers a new one in its place.[9] But if the sceptic doubts or denies (6), and if the truth of what the sceptic doubts or denies is to be a necessary condition of those doubts' making sense then Strawson would have to show that (6), a statement about the way things are, follows from (1), a statement about how we think of the world, or what makes sense to us. How could such an inference ever be justified?

Strawson's argument is this. The sceptic's doubts about the continued existence of objects make sense only if (1) is true. But it is a necessary truth that:

(2) If we think of the world as containing objective particulars in a single spatio-temporal system then we are able to identify and reidentify particulars.

And again, necessarily:

(3) If we can reidentify particulars then we have satisfiable criteria on the basis of which we can make reidentifications.

Strawson's argument actually stops here, thus showing that he regards what has been established as sufficient to imply his diagnosis of scepticism, but it is clear that it does not follow from (1)–(3) alone that objects continue to exist unperceived. The most that has been explicitly established is that if the sceptic's statement makes sense then we must have satisfiable criteria on the basis of which we can reidentify a presently observed object as numerically the same as one observed earlier, before a discontinuity in our perception of it. And this does not imply that objects continue to exist unperceived if it is possible for all reidentification statements to be false even though they are asserted on the basis of the best criteria we ever have for reidentification. Only if this is not possible will Strawson's argument be successful.

A principle that would explicitly rule out this alleged possibility would be:

(4) If we know that the best criteria we have for the reidentification of particulars have been satisfied then we know that objects continue to exist unperceived.

Either this is a suppressed premiss of Strawson's argument or it is what he means by 'criteria for reidentification of particulars' – in either case it is required for the success of his attack on scepticism. But the argument now comes down to the claim that if we think of the world as containing objective particulars then it must be possible for us to know whether objects continue to exist unperceived. We couldn't make sense of the notion of unperceived existence without having criteria of reidentification, and if we have such criteria then we can sometimes know whether objects continue to exist unperceived. I shall call this result, which is the conclusion of the argument from (1) to (4), the verification principle. If this principle is not true Strawson's argument is unsound.

It does not follow from (1)–(4) that we actually *do* know that objects continue to exist unperceived, and hence that (6) is true, but that conclusion will follow if we add to the verification principle one more premiss to the effect that:

(5) We sometimes know that the best criteria we have for the reidentification of particulars have been satisfied.

The fact that (5) is needed shows that it was wrong to interpret Strawson as making a purely deductive step from how we think, or what makes sense to us, to the way things are. (6) is not a consequence of (1) alone, but only of the conjunction of (1) and (5), and so there is an additional factual premiss which enables Strawson to make the otherwise questionable transition. And this in turn shows that Strawson was wrong to take the sceptic to be denying (6). If the truth of what the sceptic denies is a necessary condition of that denial's making sense, and if, as we've seen, it is not the case that the truth of (6) is a necessary condition of the sceptic's making sense, then the sceptic cannot be denying (6). On his grounds, to deny this would be just as unjustified as our asserting it – he argues only that our belief that objects continue to exist unperceived can never be justified.

If this is so, then the factual premiss which warrants the inference to (6) is obviously superfluous. The verification principle which the argument rests on is: if the notion of objective particulars makes sense to us then we can sometimes know certain conditions to be fulfilled, the fulfillment of which logically implies either that objects continue to exist unperceived, or that they do not. The sceptic says that we can never justify our acceptance of the proposition that objects continue to exist unperceived, but now there is a direct and conclusive answer to him. If the sceptic's claim makes sense it must be false, since if that proposition could not be known to be true or known to be false it would make no sense. This follows from the truth of the verification principle. Without this principle Strawson's argument would have no force; but with this principle the sceptic is directly and conclusively refuted, and there is no further need to go through an indirect or transcendental argument to expose his mistakes.

Strawson's apparently more complicated account of scepticism about other minds is essentially the same as this. In order for me to understand, or make sense of, talk of *my* experiences, I must at least understand the ascription of experiences to others. But it is a necessary condition of my understanding this that I be able to identify different individuals as the subjects of such ascriptions. And this in turn is possible only if the individuals in question are such that both states of consciousness and corporeal characteristics are ascribable to them. But talk of identifiable individuals of this special or unique type makes sense only if we have 'logically adequate kinds of criteria' for ascriptions of such predicates to them. Hence 'the sceptical problem does not arise' – its very statement 'involves the pretended acceptance of a conceptual scheme and at the same time the silent repudiation of one of the conditions of its existence'.[10] But what the sceptic 'repudiates' is the possibility of my knowing that there are any states of consciousness other than mine, and so Strawson's characterization of the sceptic is correct only if my possession of 'logically adequate criteria' for the other-ascription of a particular psychological state implies that it is possible for me to know certain conditions to be fulfilled, the fulfillment of which logically implies either that some particular person other than myself is in that state or that he is not. This must be either a suppressed premiss of Strawson's argument or an explanation of 'logically adequate criteria'.

As before, then, the sceptic is seen as maintaining both that (i) a particular class of propositions makes sense and that (ii) we can never know whether or not any of them are true. For Strawson the falsity of (ii) is a necessary condition for the truth of (i), and the truth of (i) is in turn required for the sceptic's claim itself to make sense. Therefore the success of Strawson's attack on both forms of scepticism depends on the truth of some version of what I have called the verification principle.

In *Self-Knowledge and Self-Identity* Shoemaker argues against the other-minds sceptic as follows.[11] A person who understands 'I am in pain' cannot utter those words sincerely and without a slip of the tongue unless he is in pain. Therefore, if it is possible to know whether another person understands the word 'pain' it must be possible to know whether another person is in pain. But the word 'pain' could not have an established meaning if it were not possible to be taught its meaning and possible for us to determine whether a person is using it correctly. Therefore to assert, as the sceptic does, that it is logically impossible for one person to know of another that he is in pain is to imply that the word 'pain' has no established meaning. But if the word 'pain' has no established meaning, then the putative statement that it is logically impossible for one person to know of another that he is in pain has no established meaning either. Therefore, either what the sceptic says has no established meaning, or it is false.

This conclusion is the same as Strawson's, but in summarizing the argument Shoemaker makes a further claim for it which appears to be mistaken. He says:

Of any sentence that appears to say that it is logically impossible to know that another person is in pain we must say either that it actually expresses no statement at all or that it expresses a statement that is necessarily false.[12]

But it does not follow from the necessity of the conditional 'if the sceptic's statement makes sense, then it is false' that the sceptic's statement is a necessary falsehood. Although Shoemaker does not go on to draw any conclusions from this summary of the argument that do not follow from the argument itself, later on he does claim that:

It is a necessary (logical, or conceptual) truth, not a contingent one, that when perceptual and memory statements are sincerely and confidently asserted, i.e., express confident beliefs, they are generally true.[13]

One argument he gives for this starts out as follows:

(I) A primary criterion for determining whether a person understands such terms as 'see' and 'remember' is whether under optimum conditions the confident claims that he makes by the use of these words are generally true.[14]

It is essential for anyone's using the words 'see' and 'remember' correctly – and hence for their having the established meanings they have – that statements made by the use of those words be generally true. Therefore, if perceptual and memory statements were not generally true then 'see' and 'remember' would not have the meanings they appear to have, and there would be no perceptual or memory statements.

To say that the words 'see' and 'remember' would not have the meanings they do have unless the statements people made by the use of those words were generally true is explicitly to rule out the possibility of our understanding those statements when they are, unknown to us, always false, or false most of the time, although they appear to be true and hence we *believe* them. Therefore this argument too depends on the truth of the verification principle. But more is needed in order to prove that it is a necessary truth that perceptual and memory statements are generally true. The most that has been established is that the putative statement that it is not the case that perceptual and memory statements are generally true is either false or meaningless. But this alone does not imply that it is a necessary falsehood, and so does not imply that it is a necessary truth that perceptual and memory statements are generally true.

The rest of the argument is:

(II) So to suppose that (a) it is only a contingent fact, which could be otherwise, that confident perceptual and memory statements are generally true is to suppose that (b) we have no way of telling whether a person understands the use of words like 'see' and 'remember', or means by them what others mean by them, that (c) we can never have any good reason for regarding any utterance made by another person as a perceptual or memory statement, and that (d) we could therefore never discover the supposedly contingent fact that perceptual and memory statements are generally true. And this is a logically absurd supposition.[15]

But the conclusion that it is a necessary truth that perceptual and memory statements are generally true does not follow from this alone because (b), (c), and (d) do not follow from (I) and (a). All that follows is that it is a contingent fact that any person understands 'see' and 'remember'. And that this is a contingent fact does not itself imply that (b) we can have no way of telling whether it obtains or that (c) we can never have any good reason for regarding any utterance as a perceptual or memory statement, since the contingency of 'p' does not in general imply that we can never find out that p. Without some independent support for this last step the argument would fail. Given (I), (c) and (d) do follow from the assumption that perceptual and memory statements *are not* generally true, but they do not follow from the quite different assumption that it is a contingent fact that perceptual and memory statements are generally true.

Shoemaker's independent argument is that in trying to discover by inductive means the allegedly contingent fact that perceptual and memory statements are generally true I could not rely on anything that I believe on the basis of observation or memory. But there is no other way in which I could come to know it, therefore I could never know it. From the assumption (shared by the sceptic) that if it is a contingent fact that p then our acceptance of 'p' can be supported only by experience or by inductive means, and the fact that we could not rely on perception or memory in order to establish that our perceptual and memory beliefs are generally true, Shoemaker concludes that it is a necessary truth that those beliefs are generally true. But this does not follow, and the most that he has shown, as he himself sometimes points out,[16] is that a conditional statement to the effect that if . . . then perceptual and memory beliefs are generally true is a necessary truth.

What should the antecedent of such a conditional be? Shoemaker says that 'it follows from the logical possibility of anyone's knowing anything about the world that perceptual and memory beliefs are generally true',[17] but this alone raises no difficulties for the sceptic who denies that we can know anything about the world. He too insists on the truth of that conditional. It is no accident that those concerned with all of our knowledge of the world concentrated on perception and, to a lesser degree, on memory.

Rather than dealing with the conditions of *knowledge*, then, those conditionals must assert that the truth of what the sceptic doubts or denies is a necessary condition of the *meaningfulness* of that doubt or denial. But even this could fail to be a conclusive refutation of the sceptic. If only a restricted class of propositions is in question it is always open to the sceptic to accept the argument and conclude that talk about, say, the continued existence of unperceived objects really doesn't make sense to us. Although he wouldn't, and needn't, say this at the outset, he would be forced into it by an argument which relied on the truth of the verification principle. Far from refuting scepticism, this would make it stronger. Not only would we be unable to know whether the proposition allegedly expressed by a certain form of words is true, we would not even understand those words.[18] A successful anti-sceptical argument will therefore have to be completely general, and deal with the necessary conditions of anything's making sense, not just with the meaningfulness of this or that restricted class of propositions.

Furthermore, it won't be enough to deal simply with all of language *as it now is*. David Pears described the conclusions of Strawson's arguments as 'conditional necessities' to the effect that such-and-such is necessary if we are to think and speak as we now do.[19] But even if such conditionals are true, it is still open to the conventionalist to claim that no 'theoretical' justification has been given for our acceptance of the propositions the sceptic doubts or denies, since we could simply give up our present ways of thinking and speaking (of which they are the necessary conditions) and adopt others (of which they are not). Transcendental arguments must yield more than 'conditional necessities' in this sense – they must make these sceptical and conventionalist replies impossible.

Kant thought that his transcendental proofs counted in a unique way against both scepticism and conventionalism because their conclusions were synthetic and could be known *a priori*. They are shown to have this status by a transcendental argument which proves that the truth of its conclusion is a necessary condition of there being any experience or thought at all. If the conclusion were not true, there could be no experience to falsify it. For Kant, proofs that such-and-such is a necessary condition of thought or experience in general therefore have a special feature which is not shared by other proofs that one thing is a necessary condition of another,[20] and because they have this feature they can answer the 'question of justification'.

Suppose we have a proof that the truth of a particular proposition S is a necessary condition of there being any meaningful language, or of anything's making sense to anyone. For brevity, I will say that the truth of S is a necessary condition of there being some language. If we had such a proof we would know that S cannot be denied truly, because it cannot be denied truly that there is some language. The existence of a language is a necessary condition of anyone's ever asserting or denying anything at all, and so if anyone denies in particular the proposition that there is some language it follows that it is true. Similarly, it is impossible to assert truly that there is no language. This suggests that there is a genuine class of propositions each member of which must be true in order for there to be any language, and which consequently cannot be denied truly by anyone, and whose negations cannot be asserted truly by anyone. Let us call this the 'privileged class'.

There are some propositions which it is impossible for one particular person ever to assert truly. For example, Descartes cannot assert truly that Descartes does not exist – his asserting it guarantees that it is false. Also, there are some propositions which it is impossible for a particular person to assert truly in a certain way, or in a particular language. I can never truly say (aloud) 'I am not now speaking', but everyone else can sometimes say this of me without falsity, and I myself can write it or think it without thereby demonstrating that it is false. Similarly, de Gaulle cannot truly say 'De Gaulle cannot construct an English sentence', but anyone else can truly say this of de Gaulle, and he himself can truly say in French that he cannot construct an English sentence. Furthermore, there are some propositions which it is impossible, not just for one person, but for any member of a particular class of people to assert truly. A Cretan cannot assert truly that every statement made by a Cretan is false – if he does assert this it must be false – but of course any non-Cretan can assert this without thereby guaranteeing its falsity. But the 'self-guaranteeing' character of the members of the privileged class is more general than that of any of these. There is no one, whoever he might be, whatever language he might speak, or whatever class of people he might belong to, who could truly deny any of the members of the privileged class of propositions.

Now no *true* proposition could be denied truly by anyone. But for any proposition S which is a member of the privileged class, the truth of S follows from the fact that somebody asserted it, or denied it, or said anything at all, and this does not hold for all true propositions generally. It might also be argued that since a necessary truth couldn't be false under any circumstances, it couldn't be denied truly under any circumstances either, and hence that all necessary truths belong to this class. This might be so, but from the fact that a proposition is a member of the privileged class it does not *follow* that it is a necessary truth, and so it seems that there are some propositions, such as 'There is some language', the truth of which is necessary for anyone's ever asserting or denying anything, but which are not themselves necessary truths.[21] It could have been, and undoubtedly was, the case at one time that there was no language, and it probably will be again. Although it could not be truly denied, still it might have been, and might yet become false.

The existence of the privileged class is obviously important, since if it could be proved that those propositions which the sceptic claims can never be adequately justified on the basis of experience are themselves members, then from the fact that what the sceptic says makes sense it would follow that those propositions are true. This would be a way of replying to the sceptic while still acknowledging the contingency of the things he questions. If those propositions could be shown to belong to the privileged class there would appear to be no more sceptical questions left open, as there are at every point when we try to answer his questions directly. In general, giving an answer to the question 'What are the necessary conditions of X?' does not tell one way or the other about the answer to the question 'Do those conditions obtain?' But in the special case of asking for the necessary conditions of there being some language, giving an answer to the first implies an affirmative answer to the second. One's asserting truly that the truth of S is a necessary condition for there being some language implies that S is true. Therefore there isn't another question about the truth-value of S yet to be answered, and anyone who denied that we know it and still demanded empirical evidence for its truth would have failed either to have understood or to have been convinced by the argument. In either case the proper reply would be to go through the argument again.

The question now arises whether there is anything special, and perhaps unique, about transcendental arguments even when they deal with the necessary conditions of language in general, or of anything's making sense. Is it only because Strawson's and Shoemaker's arguments are limited in scope that they depend on an appeal to the verification principle? There are some general reasons for being pessimistic on this question. Although it seems to me unlikely that there should be no members of the privileged class, we have yet to find a way of proving, of any particular member, that it is a member. More specifically, we have yet to show that those very propositions which the epistemological sceptic questions are themselves members of this class. It is obviously extremely difficult to prove this, and not just because talk about 'language in general' or 'the possibility of anything's making sense' is so vague that there seems to be no convincing way of deciding what it covers and what it excludes. That is certainly a difficulty, but there are others. In particular, for any candidate S, proposed as a member of the privileged class, the sceptic can always very plausibly insist that it is enough to make language possible if we *believe* that S is true, or if it looks for all the world as if it is, but that S needn't actually be true. Our having this belief would enable us to give sense to what we say, but some additional justification would still have to be given for our claim to *know* that S is true. The sceptic distinguishes between the conditions necessary for a paradigmatic or warranted (and therefore meaningful) use of an expression or statement and the conditions under which it is true.

Any opposition to scepticism on this point would have to rely on the principle that it is not possible for anything to make sense unless it is possible for us to establish whether S is true, or, alternatively, that it isn't possible for us to understand anything at all if we know only what conditions make it look for all the world as if S is true, but which are still compatible with S's falsity. The conditions for anything's making sense would have to be strong enough to include not only our beliefs about what is the case, but also the possibility of our knowing whether those beliefs are true; hence the meaning of a statement would have to be determined by what we can *know*. But to prove this would be to prove some version of the verification principle, and then the sceptic will have been directly and conclusively refuted. Therefore, even when we deal in general with the necessary conditions of there being any language at all it looks as if the use of a so-called transcendental argument to demonstrate the self-defeating character of scepticism would amount to nothing more and nothing less than an application of some version of the verification principle,[22] and if this is what a transcendental argument is then there is nothing special or unique, and certainly nothing new, about this way of attacking scepticism.

What we need to know at this point is whether or not some version of the verification principle is true. It is not my intention to discuss that issue now, but I do want to insist that it is precisely what must be discussed by many of those who look with favour on the much-heralded 'Kantian' turn in recent philosophy. It could be that we are not as far as we might think from Vienna in the 1920's.

For Kant a transcendental argument is supposed to answer the question of 'justification', and in so doing it demonstrates the 'objective validity' of certain concepts. I have taken this to mean that the concept 'X' has objective validity only if there are X's, and so demonstrating the objective validity of the concept is tantamount to demonstrating that X's actually exist. Kant thought that he could argue from the necessary conditions of thought and experience to the falsity of 'problematic idealism' and so to the actual existence of the external world of material objects, and not merely to the fact that we believe there is such a world, or that as far as we can tell there is.

An examination of some recent attempts to argue in analogous fashion suggests that, without invoking a verification principle which automatically renders superfluous any indirect argument, the most that could be proved by a consideration of the necessary conditions of language is that, for example, we must *believe* that there are material objects and other minds in order for us to be able to speak meaningfully at all. Those propositions about what we believe or about how things seem would thereby have been shown to belong to the privileged class. Although demonstrating their membership in this class would not prove that scepticism is self-defeating, it would refute a radical conventionalism of the kind outlined earlier. It would then be demonstrably false that, for every one of our present concepts, we could dispense with it and still find our experience intelligible. But until this much has been shown, not even part of the justification Kant sought for our ways of thinking will have been given.

[1] I am indebted to many friends and colleagues for their criticisms of an earlier version of this paper. I would like to thank in particular Martin Hollis and Thomas Nagel.

[2] Kant, *Critique of Pure Reason*, tr. N. Kemp Smith (London: Macmillan, 1929), A 84 ff./B 116 ff.

[3] Kant, B xxxix, footnote.

[4] When I speak of 'the sceptic' I do not mean to be referring to any person, living or dead, or even to the hypothetical upholder of a fully articulated philosophical position. I use the expression only as a convenient way of talking about those familiar philosophical doubts which it has been the aim of the theory of knowledge at least since the time of Descartes to settle.

[5] R. Carnap, 'Empiricism, Semantics and Ontology', Appendix A in his *Meaning and Necessity* (University of Chicago Press, 2nd ed. 1956).

[6] See J. O. Urmson, 'Some Questions Concerning Validity' in *Essays in Conceptual Analysis*, ed. A. Flew, (London: Macmillan, 1956), 120. According to the still fashionable view that all mathematical truths are true by virtue of the meanings of their constituent words, this assumption would also render 'absurd' all questions of the form 'Does 3695 times 1583 really equal 5849185?'. Given the meanings of the constituent words and numerals, it *follows* that the answer is 'Yes'. Has the question therefore been 'exposed' as 'absurd'?

[7] P. F. Strawson, *Individuals* (London: Methuen, 1959), 15.

[8] Ibid., 35.

[9] Ibid., 35–6.

[10] Ibid., 106.

[11] S. Shoemaker, *Self-Knowledge and Self-Identity* (Ithaca: Cornell University Press, 1963), 168–9.

[12] Shoemaker, 170.

[13] Ibid., 229.

[14] Ibid., 231.

[15] Ibid., 231–2. In this and the previous quotation I have inserted numerals and letters into Shoemaker's text.

[16] Ibid., 238.

[17] Ibid., 235.

[18] That this result follows from an application of the verification principle seems to me more an argument against the verification principle than against scepticism. Ayer expresses a somewhat similar belief in discussing Strawson. See *The Concept of a Person and Other Essays* (London: Macmillan, 1963), 110.

[19] *Philosophical Quarterly* XI (1961), 172.

[20] 'Through concepts of understanding pure reason does, indeed, establish secure principles, not however directly from concepts alone, but always only indirectly through relation of these concepts to something altogether contingent, namely, *possible experience*. When such experience (that is, something as object of possible experiences) is presupposed, these principles are indeed apodeictically certain; but in themselves, directly, they can never be known *a priori*. Thus no one can acquire insight into the proposition that everything which happens has its cause, merely from the concepts involved. It is not, therefore, a dogma, although from another point of view, namely from that of the sole field of its possible employment, that is, experience, it can be proved with complete apodeictic certainty. But though it needs proof, it should be entitled a *principle*, not a *theorem*, because it has the peculiar character that it makes possible the very experience which is its own ground of proof, and that in this experience it must always itself be presupposed.' Kant, A 737/B 765.

[21] The tendency to confuse these two different kinds of necessity has seemed an almost inevitable occupational hazard in transcendental philosophy, with its claims to establish necessary or 'conceptual' truths (cf. Shoemaker). If to say that a proposition is 'necessary' or 'conceptual' is only to say that it must be true in order for us to have certain concepts or for certain parts of our language to have the meanings they have, then it does not follow that 'necessary' or 'conceptual' truths are not contingent. Perhaps my privileged class will provide a way of keeping these different kinds of necessity distinct.

[22] This suspicion is strongly confirmed by Judith Jarvis Thomson's excellent account of the verificationism in Malcolm's argument against the possibility of a private language (*American Philosophical Quarterly* I [1964]). Stuart Hampshire's discussion of the necessary conditions for any language in which a distinction can be made between truth and falsity, while of the required generality, will have force against scepticism only if it is interpreted as resting on a verification principle (i.e., if in order for us to 'successfully identify' an X, X's must actually exist). Hampshire himself does not directly apply the argument to scepticism (*Thought and Action* [London: Chatto and Windus, 1959], ch. 1).

VII

STRAWSON ON
TRANSCENDENTAL IDEALISM

H. E. MATTHEWS

In his book *The Bounds of Sense*, Strawson argues that a distinction can be made between two 'strands' in Kant's thought in the *Critique of Pure Reason*. On the one hand, there is what Strawson calls the 'analytic' strand, in which Kant is concerned with 'the set of ideas which forms the limiting framework of all our thought about the world and experience of the world'. On the other, there is the doctrine of transcendental idealism. These two strands, Strawson maintains, are not merely distinguishable, they are independent of each other. The analytic strand contains much that is worth preserving; the doctrine of transcendental idealism, however, is incoherent and based on a misleading analogy. There is no case for preserving any part of it, and it can be abandoned without any real damage to Kant's analytic achievements.

Since Kant himself thought of transcendental idealism as his major philosophical insight and as the means of solving most of the main problems in philosophy, it is only natural to ask how far Strawson's low opinion of the doctrine is justified. Is transcendental idealism indefensible? Is it logically independent of Kant's analysis of the structure of experience? These are the questions I shall attempt to answer in this paper. Much of the paper will inevitably be concerned with the interpretation of what Kant says, in the first *Critique* and elsewhere. But my aim is not primarily to contribute to Kantian scholarship. It is more to examine a particular philosophical doctrine, which I think has certain merits and which can be considered to a certain extent without reference to the fact that Kant invented it. Only to a certain extent, however; since Kant was its inventor, his statement of its main tenets must be taken as defining what the doctrine is, and it is legitimate to require that any doctrine purporting to be 'transcendental idealism' should be consistent with at least *most* of what Kant said. On the other hand, it is possible to admit that Kant may have been guilty of inconsistencies and confusion in working out the doctrine, so that one's interpretation of it need not necessarily fit *everything* which Kant said.

From *Philosophical Quarterly* **19** (1969), pp. 204–20. Reprinted by permission of the author and *The Philosophical Quarterly*.

The validity of Strawson's criticisms of transcendental idealism obviously depends on the correctness of his interpretation of the doctrine. To begin with, therefore, I shall examine his interpretation and offer an alternative interpretation; then, in the latter part of the paper, I shall consider Strawson's main criticisms in the light of this discussion.

I

Strawson appears to see transcendental idealism as a 'two-layer' doctrine. He often refers to an element of 'a relatively familar kind of phenomenalistic idealism' in Kant's thought; and on p. 22 says that 'Kant, as transcendental idealist, is closer to Berkeley than he acknowledges'. On the other hand, Strawson seems to think of this phenomenalism as only an element in transcendental idealism: transcendental idealism proper, as it were, is the doctrine that the real world is a supersensible world of non-spatio-temporal things in themselves, which stand to each other in a mysterious quasi-causal relation which Strawson calls the 'A-relation'. The result of this relation is the representations in the human mind which are the objects of our knowledge. Graham Bird, in his book *Kant's Theory of Knowledge,* coined the name 'noumenalism' for a view not unlike this, and for the sake of brevity I shall use this name to refer to this latter element in Strawson's interpretation of Kant's doctrine. One can say, then, that for Strawson transcendental idealism equals phenomenalism plus noumenalism. If he is right, Kant in effect believed that there were two worlds, two domains containing two types of entity. One world contains 'representations' or 'appearances', which, like Berkeley's 'ideas', have a purely mental existence. The ordering of these representations in accordance with the forms of intuition and the categories produces the world of empirical reality, but what we normally call 'real objects' are in fact mere appearances, which exist only in us. The real world is supersensible; it contains things in themselves which are not in space and time and which are not knowable in the ordinary way, although their existence needs to be postulated for various reasons.

It would be foolish to deny that Kant can be interpreted in this way, and all that I hope to show in this paper is that a different version of transcendental idealism can also be put forward, that this version fits most of what Kant said, and that it avoids Strawson's criticisms. Whether or not it is Kantian (and I believe it is), this version of transcendental idealism seems to me to be of independent philosophical interest because of the light it throws on some central problems in philosophy.

The crucial question is whether Kant believed in two worlds, a sensible world of appearances in the mind and a supersensible world of reality outside the mind. The best way to approach this question is probably from the phenomenalist end, since if it can be shown that Kant did not think that the immediate objects of experience were Berkeleyan ideas in the mind, this will cast doubt on Strawson's dualistic interpretation. How close, then, was Kant to Berkeley? Kant himself, as Strawson would admit, believed they were poles spart. In the second edition 'Refutation of Idealism' (B 274 ff.), he claims to have undermined in the Transcendental Aesthetic the very ground on which Berkeley's idealism rests. The reference is to the conclusion drawn in the Aesthetic that space and the things in space were 'empirically real'. Whereas Berkeley regarded 'the things in space as merely imaginary entities'[1] because he thought of space itself as a 'nonentity' (B 274), Kant accepted that objects in space (and in time) were real, and denied that they were a 'mere illusion' (B 69). What is the issue between Kant and Berkeley here? Berkeley believed, of course, that the immaterialist could still make the distinctions we normally make between perceiving real things and dreaming, imagining, suffering from hallucinations, etc. In that sense, he held that the things in space were real and not imaginary. But since he also held that the only things which exist are minds and the ideas in them, it is clear that he has to give a different significance to the distinction from the one we normally give it. We should normally say that what is only in the mind is 'merely imaginary', and that real spatial things are things which exist independently of being perceived. If Berkeley denies that anything can exist without being perceived, then he is assimilating real things to 'imaginary entities'. So, if Kant is to differentiate his position from Berkeley's, he must show that things in space are not in the mind. In a way it is easy for Kant to do this. There is a clear distinction, he thinks, between what is in the mind and what is outside it. What is in the mind is accessible to inner sense and is in time but not in space; what is outside the mind is accessible to outer sense and is in time and also in space. How then can anyone say that spatial objects are in the mind? Spatial objects are not objects of inner sense and they are quite genuinely in space as well as in time.

This kind of move is, however, in Strawson's opinion 'disingenuous'. Its disingenuousness is manifested in the first edition version of the Fourth Paralogism (A 366–405). Kant is here arguing against the Cartesian doubt about the existence of the external world, and claims that transcendental idealism can remove the causes of this doubt. Cartesian doubt is possible, he argues, only if one starts from the assumption that the only *immediate* objects of awareness are our own mental states and that it is only of their existence that we have direct, non-inferential knowledge. The existence of objects outside the mind can only be known, if this is so, indirectly, by inferring them as causes of our mental states. This kind of inference is, of course, shaky, and can give us no grounds for certainty about the existence of matter. The answer to this kind of doubt, Kant thinks, is to deny its starting-point – that is, to deny that our knowledge of the existence of matter is any more indirect or inferential than our knowledge of our own mental states. Transcendental idealism allows us to make this denial, because it treats the objects of outer and inner sense as being on the same level, that of immediate objects of perception. Cf. A 371: 'In order to arrive at the reality of outer objects I have just as little need to resort to inference as I have in regard to the reality of the object of my inner sense, that is, in regard to the reality of my thoughts. For in both cases alike the objects are nothing but representations, the immediate perception (consciousness) of which is at the same time a sufficient proof of their reality.'

It is passages such as this one which convince Strawson that there is a strong element of Berkeleyan phenomenalism in Kant. Elsewhere in the same section Kant describes matter as 'only a species of representations' and says that space is 'in us', and this kind of language is certainly reminiscent of Berkeley. But in view of Kant's explicit distinction between his views and Berkeley's, it is worth at least asking whether this is the only construction which can be put on these remarks. What else could be meant by saying that space is 'in us' or that matter is only a species of 'representations', if not that space and spatial objects are simply ideas in our minds?

When Berkeley says that ideas are 'in our minds', he means that they only exist when perceived by a mind or minds. Cf. *Principles* Pt. 1.3: 'The table I write on I say exists, that is, I see and feel it; and if I were out of my study I should say it existed – meaning thereby that if I was in my study I might perceive it, or that some other spirit actually does perceive it.' In the argument of the *Dialogues,* he assimilates ideas to 'sensations', which of course exist only when had by some individual. The general tenor of his thought, then, is that space and spatial objects can only exist when actually perceived by some individual (human or divine). This is what leads Kant, as I have already mentioned, to accuse Berkeley of making space and spatial things 'merely imaginary entities', since it is part of the normal conception of material reality that what is real can exist unperceived, and that only the imaginary depends for its existence on our thinking about it. Since Kant rejects this view of space, it is clear that he did not attach the same meaning to saying that space was 'in us' as Berkeley did. So what meaning did he give this phrase? What point was he making by its use? The answer to these questions is made pretty plain, I think, in the Transcendental Aesthetic (A 26/B 42): 'It is, therefore, solely from the human standpoint that we can speak of space, of extended things, etc. If we depart from the subjective condition under which alone we can have outer intuition, namely, liability to be affected by objects, the representation of space stands for nothing whatsoever. This predicate can be ascribed to things only in so far as they appear to us, that is, only to objects of sensibility.' A passage from the *Critique of Practical Reason* (p. 100 footnote) is also relevant: 'Names that designate the followers of a sect have always been accompanied with much injustice; just as if one said, N is an Idealist. For although he not only admits, but even insists, that our ideas of external things have actual objects of external things corresponding to them, yet he holds that the form of the intuition does not depend on them but on the human mind.'[2] To say that space and spatial objects are 'in us', such passages suggest, is not to say that they are a particular type of thing, the type of thing which exists, as a sensation does, only in an individual mind. It is rather to say that thinking in spatial terms, thinking of things as having a position in space, as being extended in space, as having spatial relations to other things, etc., is a purely *human* way of thinking, determined by the nature of human experience (or perhaps one should say determining the nature of human experience). It is not to deny the genuineness of the distinction between what goes on in our minds and what goes on outside, but to insist that this distinction is only genuine from the point of view of human experience: other types of being might well make the distinction in a different way from us, or make no distinction at all. To put it another way, whereas Berkeley was making an *ontological* point about the types of things there are, Kant was making an *epistemological* point about the limits of human knowledge. Limits are set to what we can know by the nature of human experience: in order for a human being to have experience of the world, and hence to have empirical knowledge, his senses must be affected by things, he must see these things as in space and time, and he must be able to describe his experience in terms of certain concepts (substance, causality, etc.) which can only have a determinate content when applied to a spatio-temporal experience. If either the 'matter' (stimuli affecting the senses) or the 'form' (space, time, and the categories) were lacking, there would be something different from a human experience: either there would be a mere 'rhapsody of sensations' or there would be a mere set of statements employing concepts which could be given no empirical content. In either case, 'knowledge', as human beings understand that word, would be impossible. If experience were a mere rhapsody of unconceptualized sensations, no statements could be made about it (since making statements involves the use of concepts), and hence no *true* statements. If it were purely conceptual and had no sensory element, then statements could be made, but they could be neither verified nor falsified by human beings, and hence would have no knowledge content for them.

If this is right, then Kant does not mean by an 'appearance' or a 'representation' a particular type of *thing,* such that it only exists in the mind. To talk about 'appearance' is rather to talk about things *from a particular point of view,* namely, as they are experienced by human beings. In saying that 'objects are nothing but representations', Kant is not denying, as Berkeley would, the existence of extra-mental objects, but simply asserting that the way we experience objects, and the kind of knowledge we can have of them, depends on the nature of our human experience.

If talk about 'appearances' is talk about things from a certain point of view rather than talk about a type of thing, then so is talk about 'things in themselves' (and here we pass from Kant's alleged phenomenalism to his alleged 'noumenalism'). It is significant, in this context, that Kant often uses the expression 'things as they are in themselves', a form of words which suggests that, in distinguishing 'appearances' from 'things in themselves', he is contrasting, not two types of thing, but two ways of considering the same things. If this interpretation is correct, then 'things in themselves' will not be things located, like Locke's 'substance', as it were, 'behind' the things we perceive, but will be the very same things that we perceive, but considered from some other point of view than that of human experience.

There are various ways of considering things which might be contrasted with the standpoint of human experience. We might contrast the world as we describe it, using our conceptual framework, with the world that we thus describe, the world to which our concepts are applied. The latter world would be *ex hypothesi* indescribable and, in a sense, unthinkable. Nothing could be said in detail about it, but its existence seems to be implied in the talk of applying concepts: if we apply concepts, we presumably apply them *to* something. This is what Kant seems to mean by 'noumena' in this passage from the section on 'Phenomena and Noumena' (B 306): 'At the same time, if we entitle certain objects, as appearances, sensible entities (phenomena), then since we thus distinguish the mode in which we intuit them from the nature that belongs to them in themselves, it is implied in this distinction that we place the latter, considered in their own nature, although we do not so intuit them, or that we place other possible things, which are not objects of our senses but are thought as objects merely through the understanding, in opposition to the former, and that in so doing we entitle them intelligible entities (noumena).' To put it another way, if we want to distinguish a human standpoint on the world from other possible standpoints (non-spatio-temporal ones), then we imply that these are all standpoints on the *same* world, which is 'in itself' prior to all description and in that sense indescribable and unknowable. (I shall consider a possible objection to this line of argument later in the paper.)

Secondly, we can contrast the world as we experience it with the world as it might be experienced by a being with another type of intuition. Whereas our intuition is sensory, that of this being would be 'intellectual'. The being in question is, of course, God, as conceived in traditional Christian philosophy, the being who knows things, not in the passive way that we know them but in an active way, as their creator. In order to make the contrast in this way, it is not necessary to assume the existence of such a God, since the point of the contrast is simply to emphasize that, whatever God may be able to do, *we* (human beings) cannot know things in this way by intellectual intuition. The warning is necessary, because if we do think of our intellect as intuitive, we shall be led to assume the existence of noumena in a positive sense, an illusory assumption made by 'dogmatic' (rationalist) metaphysicians. Noumena in this sense are the objects of a non-sensory form of intuition: they are 'supersensible' in the sense that they are not accessible to our senses. Kant nowhere asserts that there are noumena in this sense, or that there is a world which is 'supersensible' or 'intelligible' in this sense of those words. He is interested only in the *concept* of such a world, as a way of emphasizing the limitations of our human knowledge and avoiding the excesses of dogmatic metaphysics.

Kant sometimes, it is true, asserts the existence of a 'supersensible' or 'intelligible' reality, but he is using these words in a different sense, corresponding to a third form of the contrast. We can draw a contrast between two different human standpoints: the standpoint of experience and the standpoint of action. That is, we may consider our environment either as something to be known about or as a field for our activity. This contrast is made by Kant in the form of a distinction between the standpoints of theoretical and practical reason. From the standpoint of theoretical reason, we are concerned to make statements which will have a certain knowledge-content, which will be verifiable or falsifiable. A statement whose truth cannot be established (in principle) will be a statement without content from this point of view, although it will still be a well-formed statement as long as it is expressed in accordance with one of the logical forms of judgement. For a well-formed statement to be verifiable and to have content, the categorial concepts employed in it must be schematized, that is, applied to a spatio-temporal manifold. From this point of view, 'objective reality' is that about which true well-formed statements can be made. It is 'objective' in the sense that their truth, and its existence, do not depend on anything about the human beings making the statements.

From the other point of view, that of practical reason, we are not concerned to make statements which will be true or false, but to make statements which express presuppositions of our activity as desiring or willing beings. These statements must be equally well-formed in accordance with the logical forms of judgement, but it is not necessary that the categorial concepts employed in them should be schematized, since considerations of verifiability are not relevant here. These statements have content, not in the sense that they embody knowledge about objects, but in the sense that they express genuine presuppositions of our practical activity. In this context 'objective reality' consists of those presuppositions which hold independently of anything about the particular human agents concerned: they as it were force themselves upon the agents independently of their own wishes.

The system of these presuppositions may be called a 'supersensible world' or an 'intelligible world', not in the sense that it consists of objects which can only be known about by non-sensory or intellectual intuition, but in the sense that its 'objects' are not objects of *knowledge* at all, but 'objectively real' presuppositions of activity. Or, to put it differently, statements made from the point of view of practical reason are not statements of fact which might be verified by non-empirical means: they are not statements of fact at all, but expressions of the presuppositions of a practical attitude to the world. In talking of a 'supersensible world' in this sense, Kant is not asserting the existence of noumena in a positive sense, a world of objects lying behind the world we see and accessible only to intellectual intuition. He is rather contrasting the point of view of practical reason, of the agent, with that of theoretical reason, of the knower.

If one ignores the fact that 'supersensible' is being used in a different sense in this sort of context, one is liable to accuse Kant, as Strawson does, of noumenalism. Strawson's picture of transcendental idealism does seem to have been distorted in this way. He thinks, for instance, that it is in the sections where Kant is most concerned with moral and religious matters that he is most liable to lapse into what I have called noumenalism. But an examination of these passages will both show the sources of Strawson's misunderstanding (as I see it), and also help to clarify the rather difficult point I have just been trying to make. We are here concerned not with Kant's personal beliefs about 'God, freedom and immortality', but with the consequences of transcendental idealism for those beliefs, as he saw them.

First of all, what bearing does transcendental idealism have on belief in God? In the relevant sections of the Transcendental Dialectic, Kant's main purpose is to show the impossibility of a 'dogmatic' theoretical proof of the existence of God. His argument hinges on the conclusion of his earlier analysis of knowledge that existential propositions must be synthetic, and cannot be analytic. The same kind of argument would also hold against dogmatic attempts to prove by theoretical reasoning the *non-existence* of God. So, from the point of view of theoretical reason, we are left in an agnostic position: we cannot assert as a fact either the existence or the non-existence of God. Here transcendental idealism steps in: it allows at least the possibility of avoiding agnosticism (which Kant thought was morally dangerous). The possibility arises because transcendental idealism treats the world we experience, the world about which we can have empirical knowledge, as a world of 'appearances', and this leaves open the possibility that there might be another, non-sensible or intelligible world in which God might be located, and to which we can have access by means of our practical reason. If the interpretation which I have given above of this sort of talk is correct, then this might be paraphrased as follows: the point of view of theoretical reason is not the only possible point of view on the world which a human being might take up; he can also adopt the standpoint of practical reason. That is, he might want to make statements which are not verifiable assertions of fact about things, but express presuppositions of his practical dealings with the world. From this point of view, the (well-formed) statement 'God exists' would not be asserted as true-or-false, as embodying a piece of knowledge about the world of a non-empirical kind, but would express such things as our reverence for the moral law, coupled with a working assumption that the universe is intelligible and a determination to live our lives *as if* being good would make us happy. The point of the phrase 'as if' is to indicate that sentences like 'God exists', although apparently of a fact-stating form, do not state facts, at least of a kind which we could establish. Other sorts of being with other types of intuition might be able to assert them as facts, but for us they merely express a presupposition of our practical attitude to the world. This is also the point of calling such beliefs matters of faith rather than knowledge, and of talk of God as merely an 'Idea' in the Kantian sense, i.e., a concept which derives its meaning for us human beings only in connection with our commitments to act or think in certain ways, and not by referring to a special type of object. This interpretation fits in with Kant's remarks in *Religion within the Limits of Reason alone*, for example with this quotation[3]: 'First, in religion, as regards the theoretical apprehension and avowal of belief, no assertorial knowledge is required (even of God's existence) . . . rather is it merely a problematical assumption (hypothesis) regarding the highest cause of things that is presupposed speculatively, yet with an eye to the object toward which our morally legislative reason bids us strive. . . . This faith needs merely *the idea of God*.' (Kant's italics.) It also fits what we know of Kant's general attitudes to religious practices – his contempt for the liturgical and ceremonial aspects of religion and for all those aspects of religion which have no direct bearing on morality but depend on factual beliefs about a supersensible reality.

The things which Kant says about God from the point of view of trans-cendental idealism, then, lend no support to a 'noumenalist' interpretation. His remarks seem to have the mainly negative function of showing the impossibility of dogmatic theology and the possibility of a moral theology (in Kant's sense of that phrase). But what he says about freedom of the will is more positive and might seem, therefore, to be more favourable to a Strawsonian interpretation. Cf. the Preface to the *Critique of Practical Reason* (p. 88): 'Freedom, however, is the only one of all the ideas of the speculative reason of which we *know* the possibility *a priori* (without, however, understanding it), because it is the condition of the moral law which we know. The ideas of God and immortality, however, are not conditions of the moral law, but only conditions of the necessary object of a will determined by this law: that is to say, conditions of the practical use of our pure reason.' Kant's belief in freedom of the will, then, is admittedly stronger than his belief in God and immortality; but I hope to show that it is not an essentially different kind of belief.

Again, one must start with the negative part of Kant's argument on this point in the Third Antinomy. He argues that there are no theoretical grounds for assuming freedom of the will. Within the world of experience, an uncaused cause cannot be found: everything we can experience is caused by some antecedent event which we can also experience. This conclusion is believed to follow from Kant's earlier analysis of the structure of our experience. Thus, from the theoretical point of view, we are committed to universal determinism. But transcendental idealism allows us a possible escape from total determinism, since it asserts that the world of experience is merely phenomenal, and this makes it possible that we might be free as noumena. Only practical reason can demonstrate the reality of this noumenal freedom. The interpretation of these remarks which I would suggest is this. We can look at the world, including our own actions and states of mind, from either a theoretical or a practical point of view. From the theoretical point of view, we are interested in our own states of mind or those of other people as possible objects of knowledge. We want, that is, to make factual assertions about them – to state the temporal relations be-tween our states of mind and physical events. A state of mind can be said to be the cause of a physical event in exactly the same sense that one physcial event can be said to be the cause of another; A is the cause of B if B follows A in accordance with a rule, and this is so whether A is a mental state or a physical event. (Conversely mental states can be caused by physical events in the same sense of 'caused'.) Kant believes himself to have proved that it is a condition of the possibility of experience (i.e., of empirical knowledge) that every event we can experience, whether mental or physical, must have a cause. The principle of causality therefore applies universally to everything we can experience. But the point of view of experience, of theoretical reason, is not the only point of view from which we can consider our mental state. We can also adopt the standpoint of practical reason, of agents rather than knowers. From this standpoint, we are concerned with our mental states not as objects of knowledge, but from the point of view of our action. We do not wish to make factual statements about mental states as *causes*, but look on them as *reasons* for acting. The logical form of judgement involved in saying A is a reason for B is the same as that involved in saying that A is the cause of B, but they have a totally different significance for us. The causal statement is a factual assertion about our mental states considered as objects of experience, and to say that A is the cause of B is to assert a certain temporal relation between A and B, because time is a universal form of our experience. But from the practical point of view we are not concerned with experience (in this sense of empirical knowledge) and so temporal relations are not relevant. To say that A is a reason for doing B is not to assert or deny any temporal relation between A and B. Time does not enter into it, since we are not concerned with the question 'What can I know about A and B?' (the theoretical question) but with the question 'What ought I to do?' or 'What would be best for me to do?' (the practical question). Both the theoretical and the practical standpoints are possible for human beings; indeed, one can even say that the theoretical presupposes the practical, since knowledge involves making judgements, applying concepts, and so on, in other words, being active.

If one interprets Kant in this way, there is no warrant for saying that he thought that 'reasons' were a special kind of event taking place in a supersensible noumenal world, and palely reflected in the appearances to inner sense. Free action, action for reasons, is 'supersensible', but in a different sense from this. It is 'supersensible' in the sense that it is not an object of knowledge, either by means of sensible intuition or by means of intellectual intuition, but an essential part of our attitude to the world as agents. The mental states which we consider as motives for action are not *different things* from mental states considered as causes, but the same things considered in a different way. The phrase 'objective reality' has also a different sense, according to whether one is thinking in terms of action or of knowledge. Cf. again the Preface to the *Critique of Practical Reason* (p. 90): 'Here first is explained the enigma of the critical philosophy, viz. how we deny objective reality to the supersensible use of the categories in speculation, and yet admit this reality with respect to the objects of pure practical reason. This must at first seem inconsistent as long as this practical use is only nominally known. But when, by a thorough analysis of it, one becomes aware that the reality spoken of does not imply any theoretical determination of the categories, and extension of our knowledge to the supersensible; but that what is meant is that in this respect an object belongs to them, because either they are contained in the necessary determination of the will *a priori*, or are inseparably connected with its object; then this inconsistency disappears, because the use we make of these concepts is different from what speculative reason requires.' Freedom is 'objectively real', not in the sense that it is an object about which true factual assertions can be made, but in the sense that its status as a presupposition of our practical attitude to the world is independent of any individual's thinking.

Kant's solution to the problem of determinism, then, is not to assert the existence of a world of supersensible things, but to stress the difference between looking on ourselves as objects of knowledge, subject to causality like all other objects of knowledge, and thinking of ourselves as agents, 'from the inside'. As agents, we are not objects of knowledge, and there is no necessity to think of ourselves in causal, or indeed in temporal, terms, since such ways of thinking only have application when we are thinking of the world as a field for knowledge. One could make this point in another way which would make it sound ironically similar to Strawson's own point against determinism in his British Academy lecture 'Freedom and Resentment'.[4] Determinism, one might say, is the result of over-intellectualizing human behaviour, including one's own behaviour, of looking at it exclusively 'from the outside', as an object of knowledge, and leaving out other, equally valid, ways of looking at actions; the problem is, not solved, but dissolved by recognizing that we are not just objects for ourselves (nor, one might add, are other people simply objects for us), but also subjects, beings who act on and react to the world. Transcendental idealism, with its recognition that human ways of thinking are just that, human ways of thinking, makes possible this kind of resolution of traditional philosophical problems. This is one reason why I said earlier that transcendental idealism was of philosophical interest in its own right, and not merely a doctrine of interest only to Kant scholars.

II

On my interpretation, then, transcendental idealism is not the same as phenomenalism or as noumenalism; still less is it a combination of the two doctrines. It is not a form of phenomenalism, since Kant's 'representations' are not Berkeleyan 'ideas' or 'sense-data' (as most modern phenomenalists think of them). It is not noumenalism, since Kant does not assert the existence of a world of objects not accessible to our senses, but perhaps accessible to other beings. Transcendental idealism, and the Copernican revolution on which it is based, can best be seen, I think, as negative moves (though these negative moves had positive consequences). Kant was reacting against transcendent metaphysics of the Leibnizian type, which relies on our making sense of a God's-eye view of reality, a view *sub specie aeternitatis*. For Leibniz, the 'real' world was timeless, and our experience of things as temporal merely a result of our confused awareness of reality. Again, for Leibniz, such a truth as 'Caesar crossed the Rubicon' was 'really' analytic, although only God could see its analyticity; if we human beings saw it as synthetic, that was simply an indication of our limitations. Doctrines like these depend on our being able to compare the human view with the divine and to apply the contrast between 'appearance' and 'reality' accordingly, and this could only be possible if it were open to us to take the God's-eye view, to experience things as God might experience them, to see the relations between things as non-temporal relations between concepts. The point of the Copernican revolution is to remind philosophers that they are not Gods; that we can know about things only what human beings can know and only in the way human beings can know things, using the human conceptual framework. The point of insisting that our experience is of 'appearances' and that we cannot know anything of 'things in themselves' is, at least in part, to remind us that the things we can say, and hence the things we can know, are determined by the nature of our human experience and that we cannot step outside the limitations of our human experience. Far from Kant's asserting that there is a supersensible world of real things, or that reality is supersensible, this is precisely what he is denying; or at least he is saying that, if there is such a world, it is unknowable as far as we are concerned and cannot have any function in our factual enquiries.

Of course, Kant does sometimes talk about a supersensible or 'intelligible' world, but I would argue that the sense of the word 'supersensible' is different here. The point Kant is making in these passages is again a negative one, but it is directed at a slightly different target. This time the target is those philosophers who fail to see the difference between a human approach to the world in 'theoretical' terms, in which we passively contemplate the facts of nature, and the 'practical', or agent's, approach, which sees the world as something to be changed in accordance with human desires or values. From the theoretical point of view, we see the world as a system of substances, which are spatially related to each other, and whose states are temporally and causally related. The notions of space and time, and such concepts as *substance* and *cause,* are, Kant argues, an essential part of our theoretical reasoning, without which no content of a factual nature could be given to our judgements. But if we are interested not in making factual judgements *about* the world, but in acting *on* the world, then the necessity for thinking in these ways disappears. The assumption of universal determinism, for instance, may be a necessary presupposition of theoretical reasoning, but that does not prevent us from thinking in non-determinist ways in our practical reasoning. Our thought still has to conform to the logical forms of judgement, since these are the forms of all thinking, theoretical and practical; but it is not necessary for these forms of judgement to be given any 'factual' content. The sense of 'supersensible' in this kind of context is determined by the context: to say that we are 'supersensibly' or 'noumenally' free (or that we may be) is to say that the question of freedom is not one that arises in our theoretical thinking, but in our practical reasoning, and that the practical attitude is as valid a human approach to the world as the theoretical.

Now that we have sketched an alternative interpretation of transcendental idealism, we can see more easily whether the major criticisms made by Strawson in his book still hold good when applied to this version. Some of Strawson's objections are clearly applicable only to the version of transcendental idealism which I have rejected. Strawson argues, for instance, that the pronoun 'we' (and related pronouns) as used by Kant does not refer to human beings existing in time and space, but to a supersensible timeless being; talk about 'our' use of concepts and so on is thus a matter of making assertions about what Ryle might call 'ghostly machinery'. This use of personal pronouns does nothing, in Strawson's opinion, to make any connection between the supersensible world and the world of human beings. But, if my interpretation of Kant is correct, there is no need for him to make any such connection, since he has not asserted the existence of any world other than that of human beings. Talking about human beings as timeless subjects and talking about them as temporal objects (beings which have a history, as Strawson puts it) is not talking about two worlds; it is talking about the one world in two different ways, from two different standpoints. The very quotation which Strawson gives (from A 546-7/B 574-5) allows of this interpretation: man can look on himself either as 'phenomenon', as an object of knowledge, to be studied by empirical means, or as a 'purely intelligible object', as a being who conceives aims and values and acts upon them. The aims and values are non-empirical in the sense that their recognition *as* aims and values cannot follow from any empirical facts about the determination of our thoughts and feelings by observable stimuli: to see them as aims and values is only possible if we are thinking of ourselves in practical, not theoretical terms. The transcendental ego is not an object, belonging to a world to which only non-sensory intuition could have access:

it is simply a man's self, the very same self which in one aspect 'has a history', but thought of in a different way, 'from the inside' rather than 'from the outside' (these metaphors probably make the point as clearly as it could be made). Similarly, 'reason', 'understanding' and so on are not names of mysterious bits of non-temporal machinery, but refer to ordinary mental operations like conceptualizing, inferring, and so on. These operations can be studied empirically like anything else; but they can also be thought of, from the point of view of a critical enquiry into the conditions of human knowledge, as the features of the human way of experiencing the world which give it its characteristic form. Much the same might be said about Kant's talk of our being 'affected' by things. He is not talking here about any mysterious 'A-relation' between things in a supersensible world, but of a perfectly ordinary causal relation between observable things and our senses, such as might be empirically studied by a scientist. But although it is in this sense an 'empirical' matter that we are affected by objects, it can also be thought of, from the transcendental point of view, as a feature which helps to define our human mode of experience, to differentiate it from the intellectual intuition which a God might have, and so on. The element of strangeness, the 'phantasmagoric quality', which Strawson finds in trans-cendental idealism, vanishes, I think, if one interprets the doctrine in the way I have suggested.

Another important criticism which Strawson makes is that Kant cannot make a significant application of the contrast between appearance and reality. Our normal way of making this distinction, Strawson correctly says, is by making use of the concepts of identity of reference and of the corrected view. 'When it is said that a thing appears to be thus-and-so, but really is not, it seems to be implied that there are two different standpoints from which it would be natural to make different and incompatible judgements about the *same* thing, and that the judgement naturally made from one of these standpoints would be, in some sense, a *correction* of the judgement naturally made from the other.' (p. 250.) All this is a perfectly acceptable account of what is involved in our normal use of terms like 'appearance' and 'reality', and I do not suppose that Kant would have quarrelled with it as such. But he might well have denied its relevance to transcendental idealism, since when he says that all human experience is of 'appearances' he is not using this word in its normal sense. One might, of course, criticize Kant on this ground alone; it certainly would have been less misleading for him to use the unfamiliar word 'phenomena' throughout, since this has no confusing associations. But, granted that he is using the word in a different way from the normal, one can hardly object on the grounds that he cannot give it significance in the way that one can give significance to the word as normally used. It does not seem to me, at least, that it is unintelligible to say that all human experience is of 'appearances' in Kant's sense; which is not necessarily to say that other beings might have a more *correct* view of things than human beings do, but simply to emphasize that the human point of view is just the *human* point of view.

A connected objection made by Strawson is that Kant violates his own 'principle of significance' in talking about things in themselves. It is important, however, to understand in what sense Kant had a 'principle of significance'. To make a significant judgement, for Kant, involved using one of the logical forms of judgement and giving this form some kind of content. What kind of content depended on whether one was making a factual (true-or-false) assertion or expressing a presupposition of practical or moral life. A judgement only had factual content if the concepts used in it could be 'schematized', that is, could be applied to a spatio-temporal manifold. No judgement about things in themselves could be given factual content for this reason, and so things in themselves were unknowable. But judgements about things in themselves could be given another kind of content, if they were taken, not as assertions of factual knowledge, but as expressions of the presuppositions of practical experience. Thus there is nothing in Kant's 'principle of significance' (which is different from the positivists' verification principle) to imply that any judgement about things we cannot experience has no meaning. The principle simply makes clear what kind of meaning the judgement must have. Nor is Kant committed to saying that a judgement which is 'meaningless' in the sense of lacking factual (or other) content is also 'meaningless' in the sense of 'logically absurd'. Thus, nothing prevents him saying that the experience of a being with intellectual intuition is conceivable, in the sense that it is not self-contradictory to say 'There is a being who experiences things without being affected by them', while still maintaining that such a statement can have no factual content for us. Strawson, in his comments on these remarks of Kant, seems to me to confuse these two points. The concept of a non-sensory experience is not one that we can give any application to, but this is surely not the same as saying that it is 'inconceivable' in the sense that it is an incoherent concept.

This also indicates how one might deal with an objection which might be made to transcendental idealism as I have interpreted it. I said earlier that part of Kant's aim was to distinguish human standpoints from non-human ones, and that this implied that both standpoints were standpoints on the *same* world. It might be objected that no meaning could be attached to the word 'same' in this context. My reply would be that it depends what one means by 'attaching meaning'. The statement 'Other beings, with other types of experience, experience the same world' can certainly not be given any factual content, since the conditions for the empirical application of the concept of 'sameness' cannot be met. But the statement is not self-contradictory, and may well have a function, that of expressing the limitations of our experience, which gives it some kind of meaning. The difficulty which is met here is one which arises whenever one tries to talk about the limits of human knowledge: one seems to make quasi-factual assertions which, as Wittgenstein puts it in the Preface to the *Tractatus,* seem to involve 'finding both sides of the limit thinkable'. The only way in which one can really present the limits of human thought is by showing the confusions and contradictions which arise when one tries to overstep the limits (as Kant does in the Antinomies). But if one does try to *state* the limits (rather than just *showing* them), then the statement, despite its factual appearance, should be interpreted as having a different function.

III

The objections I have tried to deal with seem to be the core of Strawson's contention that transcendental idealism is 'incoherent' and 'unintelligible'. In conclusion, I want to consider Strawson's other claim, that transcendental idealism is independent of the 'analytic' strand in Kant's thought. To what extent is Kant's analysis of the conditions of human experience independent of his claim that all human experience is of appearances (interpreting that claim as I have done in this paper)? I have already tried to show how they are connected in particular cases, but what is the nature of the connection in general? To answer this sort of question, I think one must consider Kant in his historical context. The point of the Copernican revolution, as I have already suggested, was to reject transcendent metaphysics of the Leibnizian type. Metaphysics up to Kant's time had failed to make any progress of the kind which had been made in natural science and mathematics (cf. the Preface to the second edition of the *Critique*). The reason was that no one had systematically concerned himself with the nature or possibility of metaphysical knowledge: speculation and theorizing about transcendent entities had been carried on without first asking whether it was possible for human beings to speculate or achieve theoretical knowledge about such matters. Just as Copernicus had made progress by thinking of the apparent movements of the planets as resulting from the fact that we view them from a certain position, so we must realize that the nature of what we can know is determined by our nature as knowing beings. We can know about things only in so far as we are affected by them; and we are bound to think about them in spatio-temporal terms and in terms of our categorial concepts. The 'analytic' strand in the *Critique* is simply the detailed working out of what an 'objective experience' or a 'factual assertion' could mean for us, human beings. Thus the connection between the analytic strand and the doctrine of transcendental idealism is this: the forms of intuition and the categories can only impose limits to knowledge because 'knowledge' means 'what *we* can know' and the forms and categories are what determine the nature of our experience and hence the limits of our empirical knowledge. The analytic procedure is only possible because 'knowledge' means 'knowledge for us' and this involves seeing the human standpoint as only one among many possible standpoints. The connection between the two 'strands' in Kant's thinking is thus very close, and it is very doubtful whether they can be separated. The analytic strand on its own would probably be little more than a piece of introspective psychology. It only becomes anything more when it is seen as analysing the necessary *conditions* of experience, and it is hard to see what sense could be given to these words other than 'the features which determine the nature of human experience'. If human experience has a 'nature', then it is surely possible to distinguish it in principle from other types of experience, and to do this is to accept transcendental idealism as I have interpreted it.

Strawson might well describe my interpretation as 'anodyne', but various things can be said in its favour. First, it is at least as well grounded in the text of Kant's writings as Strawson's. Secondly, it makes sense of Kant's position in the history of philosophy (his relation to Leibniz in particular). Thirdly, it has the merit of making transcendental idealism something more than an inexplicable aberration on the part of a great philosopher. And finally, it presents the doctrine as of continuing relevance, as a philosophical tool for dealing with the apparently perennial human tendency to theorize about matters beyond the reach of human experience, and to draw conclusions from such theories which have a bearing on the conduct of human life. It may be that this tendency is, as Kant himself suggested, an inevitable result of the nature of human thinking; in which case Kant's method of dealing with it will continue to be of philosophical importance.

---

[1] All quotations from the first *Critique* are given in Kemp Smith's English translation.

[2] All quotations from this work given in Abbott's translation (London: Longmans, 6th ed. 1909). Abbott rightly comments here 'N is clearly Kant himself.'

[3] Quotation given in Greene and Hudson's translation (New York: Harper, 1960). The page reference in this edition is 142n; in the Prussian Academy edition of Kant's works the reference is vol. VI, 153–4n.

[4] Annual Philosophical Lecture, British Academy, 1962; reprinted in his collection of essays with the same title (London: Methuen, 1974). I am not suggesting that Kant's and Strawson's solutions are identical; but there are interesting parallels which might be drawn between Kant's distinction of theoretical and practical reason and Strawson's distinction between 'objective' and 'reactive' attitudes.

VIII
SELF-KNOWLEDGE

W. H. WALSH*

§2. THE problem of self-knowledge has many interesting aspects, but it cannot be dealt with exhaustively here. We must confine ourselves to those points in it which are relevant to the empiricist-rationalist controversy. A very brief statement of the history of the problem was included in Chapter II (pp. 20–3¹), at the end of which I suggested that it was necesary to concentrate on two main questions: whether introspection is in fact the same in nature as external sensation, and whether knowledge of the self is exhausted by what we know in introspection. I will begin by summarizing the opposing answers to these questions.

Empiricists have no doubt that the answer to both questions is in the affirmative. It has been a principle with the empiricist school since Locke that there is a faculty of reflection which is precisely parallel to external sensation. Reflection is a kind of internal sensation, and the results it produces are epistemologically on the same level as those of the five outer senses. By reflection I know such things as that I am feeling angry or bored, just as I know by sensation that this is hard or that tastes sour. And just as the data of the external senses provide raw material for sciences such as physics, chemistry, biology, and physiology, so the data of inner sense provide raw material for the psychologist. Psychology is the science of the self, and reflection is its indispensable instrument. It is true that psychology is a more difficult study than physics and the rest, a science in which it is altogether harder to reach agreed conclusions; but that fact can be explained simply enough. Because every self is directly accessible to a single observer only, whereas objects in the material world can be observed by as many people as like to look,[2] the psychologist faces problems, both in collecting his material and in testing his conclusions, which are unknown in the physical sciences. But empiricists would hold that though these difficulties

From *Reason and Experience*. Oxford: Clarendon Press, 1947, pp. 191–220. Reprinted by permission of the author and Oxford University Press.

*Professor Walsh wishes me to point out that his views have changed since he wrote this. For a more recent statement of them, cf. his book *Kant's Criticism of Metaphysics* (Edinburgh University Press, 1975), §§ 31–2, 42. I have included this excerpt, however, because the position it adopts is interesting and worth taking seriously, whether its author still agrees with it or not. – ED.

have impeded the progress of psychology, they are far from showing that it is an inherently impossible discipline. The success of psychologists in devising experimental methods, and the practical results they have produced, would seem to be clear testimony to the contrary.

Empiricists combine with this positive doctrine of self-knowledge a polemic directed against any suggestion that we know ourselves in any other way than by reflection or inner sense. In the case of the classical exponents of empiricism this polemic took the form of an attack on the conception of rational psychology put forward by the Cartesian school, an attack which began, moderately enough, when Locke pointed out that we have no more grasp of the true substance of the self than we have of the true substance of the material world, and which was pressed home by Hume in a famous passage from which I have already quoted.[3] The self, for Hume, was a 'bundle of perceptions', just as external objects were bundles of perceptions; it was absurd to regard it, in the manner of Descartes, as a substantial unity whose nature could be grasped in a series of rational intuitions. In modern versions of empiricism there is less emphasis on the confutation of rational psychology, a discipline now generally abandoned (except in the official philosophy of the Roman Catholic church) as impossible. But the hostility to any non-psychological doctrine of self-knowledge survives in an interesting way in the denial that philosophy can give us any special insight into the nature of mind. The propositions actually discussed by philosophers, we are now told, include besides nonsense statements of the old-fashioned metaphysical type, many empirical hypotheses, among them some which belong properly to the science of psychology. Philosophy thus appears to afford us a kind of self-knowledge, but the impression is, in fact, entirely erroneous. For the only genuine philosophical propositions are one and all analytic: philosophers are concerned not to state or explain facts, but to elucidate the proper use of symbols. Any suggestion that there can be such a thing as philosophy of mind must accordingly be dismissed as absurd.[4]

The rationalist answer to our questions is not so easy to state clearly, if only because of internal differences within the school. In general, however, we may say that rationalists are inclined to be sceptical about the view that introspection is adequately described as a form of internal sensation; and even when they admit that it is, they do not attach any very high value to the resulting science of psychology, which they believe to give only a partial view of the mind and its workings. In addition, they lay great stress on the fact of self-consciousness, a fact which (they think) is left out of account altogether in the theories of their opponents.

To illustrate the first of these points it may be instructive to consider briefly the views of Berkeley on the subject. Berkeley's account of our knowledge of the physical world is, of course, largely empiricist: he holds that we know external objects by way of ideas furnished by our senses. But he refuses to give an analogous account of self-knowledge. The self is a spirit, and a spirit is above all an active agent. To know the self we must grasp it in its activities; but that we cannot do if we are confined to knowing it by way of ideas, which are passive and inert.

If any man shall doubt of the truth of what is here delivered, let him but reflect and try if he can frame the idea of any power or active being; and whether he has ideas of two principal powers, marked by the names *will* and *understanding*, distinct from each other as well as from a third idea of substance or being in general, with a relative notion of its supporting or being the subject of the aforesaid powers, which is signified by the name *soul* or *spirit*. This is what some hold; but so far as I can see, the words *will, understanding, mind, soul, spirit,* do not stand for different ideas, or in truth, for any idea at all, but for something which is very different from ideas, and which being an agent cannot be like unto, or represented by, any idea whatsoever.[5]

How then, we may ask, do we know the self at all, if we do not know it by way of idea? The question evidently occurred to Berkeley, and he made some attempt to answer it, particularly in the second edition of the *Principles*. Thus he added to the passage just quoted the statement that 'it must be owned at the same time, that we have some notion of soul, spirit, and the operations of the mind, such as willing, loving, hating, inasmuch as we know or understand the meaning of those words'. And elsewhere he wrote that 'I have some knowledge or notion of my mind, and its acts about ideas, inasmuch as I know or understand what is meant by those words. What I know, that I have some notion of.'[6] But this language about notions is not elaborated, and we do not know precisely what Berkeley intended by it. All we do know is that he held knowledge of spirits to be quite different from knowledge of material things, and thought this was due to the fact that the essence of spirit is activity.

The point which gives plausibility to Berkeley's doctrine is the important one that there is a sense in which we seem to know ourselves from within rather than from without: as experienc*ing* rather than experienc*ed*. In the words of Samuel Alexander, we 'enjoy' ourselves and 'contemplate' objects. All rationalist theories of self-knowledge rest in the last resort on this awareness, or supposed awareness, of the self as subject: the *cogito* of Descartes, the Hegelian philosophy of spirit, the theory of mind as pure act of Gentile, are alike grounded in it. It is thought to be a form of knowledge more direct and intimate than any other, and for that reason to provide a model of what all knowledge should be. But it must be admitted that to lay exclusive stress, as Berkeley seems to do, on this aspect of the self as activity leads to paradoxical results. For, after all, there are many situations in which we seem to be not so much 'enjoying' as 'contemplating' the self, regarding it not as subject but as object. If I sit down and try to determine my precise feelings, for instance, I make myself my own object; and the 'reflection' by which I pursue my inquiries seems to be closely akin, if not to sensation itself, at least to sense-perception. Nor is there any obvious reason why I should not advance from such perceptual knowledge to a systematic study of the self as object, i.e. elaborate a psychology of the scientific type. But once that is admitted we seem to be saying that there is a good deal in the empiricist account of self-knowledge after all.

In actual fact most rationalists try, in these circumstances, to have the best of both worlds. Because of the existence of self-consciousness they are reluctant to identify introspection as such with inner sense, or, if the identification is made, to allow that inner sense is the only direct source of self-knowledge. But everyday experience constrains them to admit that the self can be an object to itself, and thus that the psychology of the empiricists is a legitimate undertaking. They qualify their approval of it, however, by maintaining that it is an 'abstract' form of knowledge; that, in considering content apart from act, the psychologist necessarily distorts his subject-matter; and that his results must accordingly be supplemented and corrected by self-knowledge obtained from within. And if they are asked where that knowledge is to be found, they refer to history, polite literature, and the writings of their fellow philosophers.

§3. The aspect of the self on which rationalists lay stress is, as will be apparent from the above, its being a subject aware of its own activities. The self is essentially an active being, and its true nature can be grasped only in self-consciousness. Against this empiricists make the point that in introspection and systematic self-knowledge (i.e. in psychology as it is normally understood) the self we are knowing is an object like any other. Now there seems no reason why this controversy should not be settled to the satisfaction of both parties, if only it can be agreed that the self is to be viewed as subject and object at once. Each side would then be right in what it affirmed and wrong only in what it denied. A solution on these lines was attempted by Kant, whose views on the matter I now propose to examine. It may be remarked in advance that his solution has, in fact, pleased neither party, and that it contains certain obvious difficulties of its own. Nevertheless, a study of it seems essential to any discussion of self-knowledge.

The following passage from Kant's lectures on 'Anthropology' (popular psychology) is perhaps as full and clear a statement of his position as can be found:

Distinguishing as reflection the inner act of spontaneity which makes possible a concept or thought from apprehension, the receptivity which makes perception, i.e.

empirical intuition, possible, and considering both acts as conscious, we can divide consciousness of ourselves (apperception) into consciousness of reflection and consciousness of apprehension. The first is a consciousness of understanding, the second of inner sense; the first is pure, the second empirical apperception (it is wrong to call it inner sense). We investigate ourselves (*a*) in psychology, using as material the representations of our inner sense; (*b*) in logic, using what intellectual consciousness provides. This seems to make the self double, which would be contradictory: (I) there is the self as subject of thinking (in logic), i.e. there is pure apperception (the merely reflective self). About this we can say no more than that it is a quite simple representation. (2) There is also the self as the object of perception, i.e. of inner sense. This self contains a plurality of determinations, through which an inner experience is made possible.

The question may be asked whether, in view of the variety of changes of mental state, changes in what he remembers or in the principles which he accepts, a man can be conscious of these changes and still say that he remains the same man (has the same soul). The question is absurd, since consciousness of such changes is only possible on the supposition that he consider himself in his different states as one and the same subject. The self is therefore formally but not materially double: double with reference to the way in which it is represented, but single in regard to its content.[7]

Whatever the difficulties of this passage, it is clear that Kant is trying in it to do justice to both the empiricist and the rationalist accounts of self-knowledge.

(A) His doctrine of inner sense, and his view of psychology, are both fundamentally empiricist. The self, he affirms, can be known as an object, and it is the business of the psychologist to investigate this aspect of it. Inner sense, the basis of psychological knowledge, is precisely parallel to outer sense, which provides the raw material for those sciences which deal with the external world. This is exactly the view of Locke. It is true that Kant goes beyond Locke by distinguishing empirical apperception from inner sense; but the distinction, unless I am mistaken, is only parallel to his general separation of sensation from sense-perception. I derive the raw material for knowledge of myself as an object from inner sense, but it is only in empirical self-consciousness that I can know my feelings precisely. It is true, too, that Kant sometimes expresses considerable scepticism about the prospects of a scientific psychology, and that in the preface to *Metaphysische Anfangsgründe der Naturwissenschaft*, where the whole subject of what constitutes a science is discussed, he contrasts psychology most unfavourably with physics, ending by suggesting that the most we can hope to get out of it is a natural history of the human mind.[8] But the objections he there brings up against psychology – that mathematics cannot be properly applied to psychological data, that we cannot carry out experiments with them because they do not exist in separation, and that they are inevitably distorted in the act of introspection – do not spring from the systematic hostility to the whole idea of such a discipline which we find in some rationalist writers. They reflect rather a conservatism which is common in scientifically minded persons, and which has its parallel today in the attitude many scientists take up to the results of psychical research.[9]

(B) But if Kant agreed with the empiricists that we can have knowledge of ourselves through inner sense, he also agreed with the rationalists that that does not end the matter. Besides the object self we must recognize the self as subject; and this, as the rationalists saw, must be conceived in terms of spontaneity or activity. The second paragraph of our quotation from the *Anthropology* shows the sort of argument Kant himself produced in support of a subject self: it is a presupposition of our having any sort of diversified experience that the self for which the experience exists should be thought of as a subject persisting unchanged through it. And this is, of course, in line with the general doctrine of the Kantian philosophy that the reality we know in everyday experience and investigate in history and the sciences stands in essential relation to a self which elaborates it in judgement. The notion of the self as an active agent plays as important a part in the Kantian scheme of things as it does in the metaphysics of Berkeley.

What is interesting about Kant's view, however, is not so much his admission of a subject self as his conception of its nature and of our knowledge of it, for it is here that he makes his distinctive contribution to the problem. Experience, he says, certainly involves the notion of a unitary self; but it is wrong to think of that self as having a content of its own, or as knowable by some form of rational intuition. To explain how knowledge of objects is possible the subject self we require need not be conceived as a substantial entity at all: it is sufficient to think of it as a formal unity preceding experience as its *a priori* condition and capable of knowing its own identity in its different acts. What we require, in fact, is the notion of a unity of apperception which is (*a*) transcendental, and (*b*) synthetic.

Few ideas in the history of philosophy have caused so much difficulty as Kant's idea of the transcendental unity of apperception; yet the thought behind the tortuous language in which it is expressed is comparatively simple. It is a presupposition of experience, as we have seen, that every object should relate to a subject, and that this subject should be aware of its own identity. But, Kant argues, we have no right to claim that we know such a subject as it is in itself, since in fact all we know of it is that it is what is formally identical in our various acts of cognition. If we abstract from these acts of cognition, taking away the content which the senses, external and internal, provide, we are left with nothing but a bare unity of consciousness, i.e. with no content at all. That is why Kant says, in the passage quoted above, that all we can say of the subject self is that it is a 'quite simple representation'. To think we can go beyond this – to think we can qualify our subject as something more than an activity which gives unity to all our knowledge – is to fall into the paralogisms of rational psychology, the futile quest for a rational doctrine of the soul which exercised Descartes and his followers. The self can and must be *thought* as subject, but it can only be *known* as object; and for that we require data from inner sense.

This point about the subject self being not so much known as thought is obviously a crucial one for our purposes, and deserves careful consideration. Before going into it, however, we must mention a complication so far passed over in silence: the distinction Kant draws between the 'real' and the phenomenal self. The distinction is important both for Kant's theory of knowledge and for his moral philosophy. It arises in theory of knowledge from Kant's desire to maintain a parallel between internal and external sensation. In external sensation, he holds, we are acquainted not with things as they exist in themselves but rather with such things as they affect our sensibilities, i.e. with appearances. Similarly, we must say that in internal sensation we intuit ourselves not as we are but only as we appear to ourselves. This gives a straight distinction between the self as phenomenon (as object of inner sense) and the noumenal reality which lies, or is thought to lie, behind that appearance; and the distinction is widely drawn on by Kant in his moral philosophy, more particularly of course in his discussion of the problem of freedom.

Now it seems natural, once we make this distinction between the 'real' and the phenomenal selves, to hold that, while the phenomenal self is known in introspection and investigated in psychology, there is also awareness of the 'real' self in self-consciousness. This is the line taken by some of Kant's successors, and it obviously fits in admirably with the requirements of the rationalist theory. But we should notice at once that Kant will have nothing to do with the suggestion. 'In the synthetic original unity of apperception', he says in a well-known passage,[10] 'I am conscious of myself neither as I appear to myself, nor as I am in myself; I am conscious only that I am. This representation is a thought, not an intuition.' He goes on to explain the point in a note, in which he argues that while apperception involves the existence of a self, it does not itself present that existence in determinate form. I am conscious of myself as an active being, and for that reason am entitled to call myself an intelligence; but to *know* myself in that capacity I should require a special intuition, 'giving me acquaintance with the determining element in me, of whose spontaneity I am conscious, prior to the act of determination, just as time [the form of inner sense] presents me with the determinable element prior to the act of determination'. Unhappily no such intuition is forthcoming, and I must conclude that my existence is determinable only through inner sense, i.e. as appearance. The same doctrine appears in the Paralogisms, though the wording there is somewhat different. The 'I think', we are told, which accompanies all the contents of my consciousness and itself represents the fact that I am a self-conscious being, 'expresses an undetermined empirical intuition, i.e. perception. . . . An undetermined perception means here only that something real has been given, given to thought in general, not however as appearance, nor yet as thing in itself (noumenon), but as something which in fact exists and is indicated as such in the proposition "I think".'[11] In other words, the fact of self-consciousness is itself enough to establish the existence of a self: 'I am' is already contained in 'I think'. But the self whose existence is 'indicated' by the 'I think' is not thereby known determinately, as Descartes supposed: the existence we ascribe to it is not existence as a category. It is a problem rather than a true datum, a problem which can be solved only if we have recourse to inner sense.

Kant puts his main position in the same context by saying that the 'I think' is always an empirical proposition, though he adds that the phrase should not be misunderstood. The unity of apperception is, of course, the ultimate presupposition of experience, and it is absurd to regard it as empirical. Again, the 'I' in the proposition 'I think' is to be taken as purely intellectual, since it belongs to thought in general. But 'without some empirical representation providing the matter for thought the act "I think" would just not take place':[12] the intellectual faculty comes into operation only on the occasion of experience. This is, of course, Kant's consistent doctrine, and it serves to explain his whole conception of the subject self. By the subject of thinking he means no more than the faculty of thought itself. Thought, as we have seen in previous chapters,[13] is for Kant an empty form without content of its own: it is to be conceived in terms of activity, spontaneity, or function, but can lead to results only if brought to bear on data which some other faculty provides. And this account applies to all intellectual acts and all forms of thinking: not only to empirical thinking, which must obviously start from what is given in experience, but to *a priori* thinking too. It is not surprising, in these circumstances, that Kant describes the subject self, the supreme condition of all thinking, as a 'mere form of consciousness', capable of accompanying the representations of inner and outer sense,[14] that he denies that it is identical with the 'I in itself', and that he maintains that we cannot know but only think it. His view is that it is not a substantial entity at all, but merely a separate aspect of the reality which we know from another point of view (as object) in inner sense. An intelligence like ours necessarily conceives the self in this dual way, as subject and object, but it does not follow that it is 'really' double. An intuitive intelligence would make no such distinction.

§4. Kant's account of self-knowledge has met with severe criticism. It is attacked in the first place by empiricists who propose to define the self, as Hume did, entirely in terms of 'perceptions'. On this view there is no subject self, and the problem of what we can know of it does not arise. Nor is it thought, by those who support this analysis, that the phenomenon of self-consciousness is of special interest to the philosopher. In Mr Ayer's words, 'all that is involved in self-consciousness is the ability of a self to remember some of its earlier states. And to say that a self A is able to remember some of its earlier states is to say merely that some of the sense-experiences which constitute A contain memory images which correspond to sense-contents which have previously occurred in the sense-history of A.'[15]

I do not propose to discuss this empiricist criticism at length here. I will say only that it seems to ignore the obvious fact that, when introspection takes place, there is a self which is introspected and a self which introspects. To say that all that happens in such a case is that one set of sense-contents confronts another is, at the least, very paradoxical. Nor do I find the equation of self-consciousness with memory at all plausible. It might, indeed, be held that the contents of inner sense, or if we like the objects of *empirical* apperception, include our own past acts of awareness.[16] But there is a world of difference between empirical and pure self-consciousness, and it is in the second that the philosopher is interested. What he wants to account for is the direct awareness of our own activity which we appear to have when we are actually performing a mental act, not the reflective awareness we can get if we give our attention to the act a moment later.

It seems strange that empiricists are not generally inclined to look on the Kantian theory of self-knowledge with a more kindly eye, since it is clearly more favourable to their point of view than it is to that of the rationalists. Kant is just as severe on the notion of rational psychology as is Hume, and his careful exposé of the pretentions and paralogisms of that 'science' has, in fact, proved quite unanswerable. Kant's subject self is, most emphatically, not a 'pure ego', yet it contrives to do justice to those facts which lead rationalist philosophers to invent 'pure ego' theories. Why then, we may ask, is it not taken more seriously by empiricists? Partly, no doubt, the reason is to be found in the unfamiliar and, indeed, repulsive terminology which Kant adopts, here as elsewhere. Expressions like 'transcendental unity of apperception' have an air of mystery about them, and one suspects that some empiricists, feeling rightly that philosophical theories ought to be expressed in clear language, are inclined to ignore or unduly neglect those which are not. But there is a more important reason. The Kantian subject self, despite its apparent harmlessness, is looked on by many empiricists as the thin end of the rationalist wedge; and confirmation is found for this interpretation of it in the fact that it was so treated by Kant's immediate successors. The unity of apperception, regarded by Kant as a mere form without content of its own, became a very different thing when criticized and developed by Fichte and Hegel, and one can scarcely expect empiricists to welcome the change. If it is true that Kant's doctrine is capable of the extravagant interpretation put upon it by his immediate successors, one cannot really be surprised at the common empiricist attitude to it. Whether it is true is what we must now attempt to determine.

I propose to deal briefly with three rationalist criticisms of Kant's account of self-knowledge: (1) that the 'I' of apperception cannot be regarded as an empty form without content of its own; (2) that Kant's so regarding it depends on his erroneous belief that the real self which is the subject of knowledge must, if we are to know it at all, itself be determinable as an object of knowledge, and (3) that the Kantian theory must be wrong because it fails to provide any explanation of the peculiar self-knowledge we have in philosophy.

1. The first of these criticisms is in line with views we have considered more than once in this book already.[17] The 'I' of apperception is looked on by Kant as an expression of that spontaneity in virtue of which I am entitled to call myself an intelligence. It is because I am a thinking being that I am capable of self-consciousness. But the faculty of thinking cannot, in human beings, lead to knowledge, since it is in no sense intuitive: it needs data given from without to work upon, and its essential function is to impose form on such material. Now this whole account of thinking has been strongly disputed, as we have seen. It was held by Hegel, for instance, that thinking cannot be thus regarded as a purely formal, discursive activity, but must rather be supposed to contain an intuitive element of its own, an element whose importance is increasingly apparent the more we give our attention to the higher forms of thought. And it is in accordance with this doctrine that Hegel refuses to regard the unity of apperception, the condition of all thinking, as nothing but an empty form. Even in Kant the unity of apperception is thought of as in some sense the source of the categories, and is described as synthetic just because of that fact. It is Hegel's contention that we cannot distinguish sharply, as Kant does, between the material of knowledge and the form imposed on it by the knowing mind. And the implication of his position seems to be that the 'I' of apperception is the source not only of the categories but indeed of the material of knowledge, too, an implication which, if not explicitly drawn by Hegel himself, is to be found quite plainly in the writings of his modern follower Gentile.

About this I can say only what I have said before, that it rests on a confusion between human thinking and the thinking we may suppose to characterize an intuitive understanding. Such an intelligence would indeed produce the materials of knowledge from itself, and impose form upon them in so doing. But that our understanding is not of this nature is clear from many considerations. To describe human experience we need, in Kantian language, concepts and intuitions; but there is no one faculty from which we can get them both. Our knowing thus appears to involve a dualism which no amount of reflection can transcend. Hegel, of course, did try hard to overcome this dualism, attempting to meet it by such devices as his theory of concrete universals and his conception of reason as a speculative faculty distinct from understanding. But just as these doctrines can clearly not be derived from anything asserted by Kant, so it cannot be shown that the Hegelian view of self-consciousness is a necessary development of the Kantian. Indeed, the two philosophers, in this as in so many other matters, stand poles apart from each other.

2. The second criticism of Kant may be thought to depend on the first, and to fail if the first fails. Yet it is perhaps worth some independent consideration. Hegel is always saying that what was wrong with the rational psychology of the pre-Kantians was that they took the self as a thing (i.e. as an object), and tried to discover its proper predicates.

But if we propose to think the mind, we must not be quite so shy of its special phenomena. Mind is essentially active in the same sense as the Schoolmen said that God is 'absolute actuosity'. But if the mind is active it must as it were utter itself. It is wrong therefore to take the mind for a processless *ens*, as did the old metaphysic which divided the processless inward life of the mind from its outward life. The mind, of all things, must be looked at in its concrete actuality, in its energy; and in such a way that its manifestations are seen to be determined by its inward force.[18]

Now it may be argued that the mistake of the old metaphysicians was repeated by Kant, despite his exposure of the paralogisms of rational psychology. To know myself as active, Kant says in a passage already quoted, I should need a special intuition of myself *qua* determining, just as I need a special intuition of myself *qua* determined to know myself as object. This suggests that he tended to think of the two sorts of knowledge as identical in kind, i.e. that he believed that to know the self as subject we should have to have some means of making it an object. And the impression is confirmed by the passages[19] where Kant argues that we cannot know the self which thinks in us since the only concepts we might invoke for the purpose are the categories, and these, since they spring from the unity of apperception, cannot themselves be used to determine it.

The impression is, in fact, a perfectly just one. Kant did hold, officially at least, that there was only one kind of knowledge, namely, knowledge of objects. To know the subject self we should have to make it an object, and this, of course, we cannot do. The most we can do is think the self as subject, not know it. And this, we may be told, is a definite defect in Kant: if the self is an essentially active agent, we must try to know it from the inside. But Kant's point is that the immediate awareness we have of the subject self in pure self-consciousness is not sufficent to give us knowledge of it. To be that it would have to be a form of intellectual intuition, which it plainly is not. And the onus of proof here is surely not with Kant but with his critics.

It may be objected that, despite his theory, Kant did make some attempt to characterize the self as an agent: in his moral philosophy. There, as is notorious, he uses the antithesis between the 'real' and the phenomenal selves to which we have drawn attention, and contrives to say a good deal about both. But it may be questioned whether he would, even there, claim to be offering *knowledge* of the 'real' self. The main aims of Kant's moral philosophy are two: first, to provide an analysis of the nature of moral action, bringing out its special character and distinguishing it clearly from all other action; and second, to explain under what conditions we might regard ourselves as moral agents. The first of these involves nothing more than, in Kant's own words, 'a development of the universally received notion of morality',[20] and does not so much as presume that a moral action has ever been done. Moral action, we are told, would, if possible, involve determination of the will by pure reason; but whether it is in fact possible is not in question. It is only in the second part of his moral philosophy, where he discusses the concept of freedom and asks under what conditions we can regard ourselves as free, that Kant makes extensive use of the contrast between the two sorts of self, putting forward his famous theory that though phenomenally determined we are noumenally free. But even here he does not say that we *know* ourselves as noumena: to make moral action conceivable we have only to *think* ourselves as free of the conditions of sensibility. That is indeed why Kant's solution of the problem of freedom is, in effect, that the problem is insoluble. It would not be so if we had insight into our 'real' selves, but moral philosophy gives us no more insight of that kind than any other inquiry. It rests on the deliverances of the moral consciousness, which are accessible to reflection and in no other way.

3. Now it may be held that this explanation is plausible only because of an important ambiguity in the word 'reflection'. 'Reflection' may be used, as it was by Locke and as it has been earlier in this chapter, as a general synonym for introspection, with the implication that it is nothing more than a kind of internal sensation (or internal sense-perception). But it may also be used to refer to the act by which reason reflects on its own operations and obtains knowledge of them. Reflection in this sense seems to be indispensable to

philosophy, or at least to parts of it such as theory of knowledge and logic. But how it could be regarded as a source of cognition on premises of the Kantian type is not by any means obvious.

We come here to the last and most difficult of the three objections which were to be considered. It raises a problem which is by no means peculiar to the philosophy of Kant, but is urgent for all philosophies of the empiricist type too. Every philosophy must explain, among other things, its own existence; or, to put it another way, among the types of knowledge which the philosopher has to consider is philosophical knowledge. A theory which produces what purports to be a comprehensive account of knowledge and yet says nothing about the status of its own propositions cannot claim to be complete. Now rationalists have no special difficulty with this problem, because they identify knowing as such with intellectual intuiting: that reason should be able to intuit its own nature seems to them far from surprising. But this solution is not, of course, open to empiricists, nor, one would suppose, to Kantians either, since both deny that intellectual intuition is possible for human beings. Yet if we refuse to admit that reason can have any special awareness of its own operations, it looks as if we must assert the fundamental propositions of theory of knowledge, for instance, to be propositions in psychology – a result which few philosophers would welcome.

I propose to confine myself here to the problem as it arises for Kant, reserving some more general remarks for the next section. Kant has a good deal to say about the nature and aims of philosophy, but he is silent, as we shall see, at the crucial point. His views can be summarized as follows:[21]

1. Philosophical knowledge is all rational knowledge, i.e. it is cognition by reason, where 'reason' is understood in a general sense. There are two sorts of philosophical knowledge, pure and applied. Pure philosophy is made up entirely of *a priori* propositions, and sets out what reason can attain when taken entirely on its own account. Applied philosophy does not confine itself to investigating *a priori* principles for their own sake; its business is to examine the concrete world in the light of such principles. Applied philosophy amounts, in fact, to the study of certain empirical questions from a philosophical point of view, and doubtless owes its place in Kant's system to the tradition that philosophy is an all-embracing discipline. It is pure philosophy which gives life to it, and on which we must concentrate our attention.

2. Pure philosophy divides into three parts, metaphysics, ethics, and logic, and each of these is made up, as we have seen, entirely of *a priori* propositions. Now if we ask Kant how reason can have *a priori* knowledge in metaphysics his answer is not in doubt. Metaphysics in the only sense in

which it can be regarded as a legitimate inquiry is made up of principles which reason prescribes to experience, and indeed must prescribe if objective knowledge is to be possible. Principles of this sort, though not derivable from sensation, have none the less an essential relation to sense-experience. Nor does the synthetic *a priori* knowledge we have in ethics present any major difficulty to Kant. Ethics deals not with statements of fact but with assertions of what ought to be; and the reason with which it is concerned is accordingly practical reason. In Kant's view it is quite wrong to think that ethics gives us any theoretical knowledge of ourselves. But what of logic? In the passage from the *Anthropology* quoted on pages 153–4 above we have seen Kant saying that in logic we know ourselves 'using as material what intellectual consciousness provides'; whilst in the lectures on logic published in Kant's lifetime by Jäsche we find logic described as the 'self-knowledge of reason and understanding'.[22] Does not this suggest that in logic we have to deal with a form of self-awareness of a special kind, and if so how is Kant to account for it? Should he say, in particular, that logic enables us to know ourselves as we actually are, in contrast to psychology, which tells us only how we appear to ourselves? That he ought to have drawn this conclusion was certainly urged by Kant's successors, and we well may think their point here a sound one.

Now Kant himself was quite clear that logic does *not* give us knowledge of our 'real' selves. The business of the logican, he held, was to reflect on concrete examples of thinking, and elicit the rules to which it conforms when true to its own nature. In formal logic we are concerned with the most general of all such rules: with the laws that govern correct thinking of any kind. But it is wrong to suppose that, when we discover these rules, we are knowing ourselves as noumena. All we are doing is bringing out a certain truth about the subject of thinking: that it must, if it is to think at all, comply with certain formal principles. But the subject of thinking, as we have seen many times already, is not for Kant a substantial self. It is only a formal unity to which all our knowledge must relate, a mere aspect of the concrete self we know in introspection. To put logic alongside psychology as a source of knowledge of the self is thus, at the least, highly misleading.

Yet, even if we accept the principles of this defence, we may well ask whether Kant is entitled to his main assumption. It seems undeniable that he thought that, in logic, reason does have insight into its own nature. He believed the laws set out by the logician to be involved in discursive thinking as such; that human reason, if it is to think soundly, must think in accordance with these and no other laws. The suggestion, favoured by some later writers on the subject, that the laws of logic are true only by definition or arbitrary convention, does not seem to have been even considered by him.

The alternative to that view which Kant accepts is to say that logical laws express necessary truths about the nature of human reason (or perhaps about discursive reason as such); but how is he to explain our knowledge of such truths? He can only say that we know them because we *see* them to be involved in all thinking. If such 'seeing' is not intellectual intuiting, what is it?

Nor is this all: if we leave the consideration of formal and pass to that of transcendental logic, further difficulties arise. Transcendental logic has to deal with our *a priori* thinking of objects, and it forms part of that propaedeutic to a true metaphysics which Kant calls 'critical philosophy'. But though Kant thought hard and long about how to validate the propositions which transcendental logic establishes, propositions such as the general law of causality, he does not appear to have given any thought to the question how it is itself possible. That there could be such an inquiry as critical philosophy, that reason could reflect on its own operations and discover the limits of its proper employment, are things which he just takes for granted. Here again he appears to have assumed that reason in a broad sense is capable of intuiting its own nature, and it is upon such an intuition that he would presumably base such a fundamental proposition as that sense and thought must co-operate in human knowledge. But it is not clear that we can admit both the intuition and the proposition it is alleged to establish: one of the two must go, yet both seem to be indispensable to Kant's thinking.

We must conclude that Kant has not sufficiently considered the sense in which some parts of philosophy seem to depend on our ascribing to reason a power of self-awareness, in virtue of which it is able to discover truths about its own nature. He recognizes, indeed, as the passages we have quoted show, that logic is, in a broad sense, a form of self-knowledge; and he should have seen that the same is true of critical philosophy itself. But his insistence on the important point that the self we illuminate in logic and theory of knowledge is really only the subject of thinking, and therefore not a substantial entity, causes him to overlook the fact that, in assuming that reason can know itself at all, he is breaking with the main principle of his philosophy. In so far as this is true, we must admit that the Kantian philosophy is self-contradictory; and there are some who would give it up just because of that fact. But it is perhaps fairer to say that Kant leaves us with a puzzle: the puzzle of knowing in what sense philosophy can throw light on the nature of mind. That there is a sense in which the study of some philosophical disciplines adds to our grasp of the nature of the self is hard to deny; but what that sense is, and how philosophical knowledge of mind relates to psychological, remain to be determined.

§5. At this point it may be helpful to put together what seem to be the main lessons of the foregoing discussion.

(A) There can be little doubt that a satisfactory account of the self must do justice to its dual character as both subject and object. What has been called the paradox of self-knowledge arises from the fact that 'we' are said to intuit 'ourselves': a proposition which suggests that the self is at once united and divided. The paradox has proved a source of inspiration to exponents of the Hegelian philosophy, with their dogma that identity is nowhere to be found apart from diversity; but even plain men must recognize that it has a certain foundation in fact. Whatever explanation we give of the duality which characterizes our conception of the self (and it may be, as Kant said, that no explanation of it is possible), conceive it as dual we must. To try to supress the subject self altogether, as some empiricists would like to do, is as absurd as to attempt to characterize the self as pure spontaneity, and deny that we can know it as object at all.

(B) The strength of the empiricist view of self-knowledge lies in the undoubted fact that we can make the self an object of knowledge, and that when we do so our experience of it is in some ways akin to our experience of external objects. In both cases we have to deal in the first place with a manifold of intuition or stream of feelings, which subsequent reflection transforms into a more or less orderly world. Our acquaintance with our own inner states is sufficiently like our acquaintance with external reality to justify us in calling it inner sensation. And just as the data of the external senses form the raw material for sciences like physics, so do those of internal sensation provide the basis for a science of psychology. This does not mean, of course, that the psychologist must conceive the self precisely on the lines on which physicists conceive their objects: clearly there are fundamental differences between the two. The phenomena of the self are not localized in space; and again they display a unity and a continuity which are without parallel in the world of inanimate objects: in their investigation we cannot (or should not) lose sight of the fact that they belong together as constituents of a single person. But though this is no doubt important if we are considering the methods the psychologist should follow in his studies, it casts no doubt on the legitimacy of his whole undertaking.

(C) The strength of the rationalist conception of self-knowledge lies in our being subjects as well as objects of introspection. That there is a sense in which we know ourselves from within, 'enjoy' ourselves as active agents, seems hard to deny. It is when we attempt to cash in on this fact, if a crude metaphor can be forgiven, that difficulties arise. What are we to make of the active self, and how are we to know it? The first answer to these questions was that given by the old-fashioned metaphysical psychology, which claimed that we could know by rational intuition the properties of the active self, which thus turned into the soul of theological discourse. The fallacies of this point of view were exposed once and for all by Hume and Kant, whose criticisms were endorsed here by Hegel. The self as subject was not to be thought of as a superior kind of object. What then was it? Kant himself made out that it was nothing more than the transcendental unity of apperception, a mere formal identity which accompanies all our thinking and as it were stamps it all as ours. About such an entity (if entity it should be called) nothing significant can be said: it is the absolute subject which admits of no predicates. But this view is not one which has carried complete conviction. Kant seems to have overlooked (or understressed) the fact that his own philosophy presupposes in human reason a power of self-awareness, and that thanks to this power we do seem able to formulate certain important truths about the knowing subject. The criticisms of his successors, in particular of Hegel, who strove to interpret the self in terms of pure spontaneity, are by no means all justified; yet over this point at least they seem to be well-founded. Philosophy does appear to shed light on the nature of the self, and Kant has not accounted satisfactorily for its doing so.

The problem of self-knowledge is thus, as we saw at the end of the last section, in essentials a problem about the relation between the knowledge of the self provided by psychology and the self-knowledge, if any, which philosophy provides. It can be put, if we like, in the question how far we can draw a legitimate distinction between introspection, where that term is taken to signify a kind of internal sense-perception, and what might be called rational reflection. It is closely bound up with the questions what philosophy aims at and how philosophical knowledge is itself possible.

Let us take two representative views.

(I) Modern empiricists, confronted with the sort of argument we have considered in this chapter, would deny that there is any problem on the ground that philosophical judgements are all analytic. Philosophy, they say, tells us nothing about the world of fact. Every factual statement is capable of translation, in principle, into sentences which record actual or possible sense-experiences or introspections; and all such statements are the object of some science or other. Philosophy is like grammar in so far as its concern is not with the factual content of sentences, but with the formal relations of the symbols they contain. There can accordingly be no clash between philosophy and the sciences, and so no possibility of there being a philosophical knowledge of mind to supplement or rival that provided by psychology.

The theory is worked out in a series of arguments which are now generally familiar. With some of these – for example, with those which profess to demonstrate the impossibility of metaphysics – we need not concern ourselves here. We should, however, notice the view of logic taken by the school. Logic is said, as it was by Kant, to be a purely formal science; but the forms which it investigates are not specifically forms of thinking. The logician's business is to examine the types of relation which hold between propositions, in virtue of which we are justified in passing from one to another.[23] Given any proposition or set of propositions, we can pass by logical inference to further propositions; and what the logician does is to investigate the principles on which our argument proceeds. But there is no single set of principles to which we are bound in formal reasoning, no group of logical rules which constitutes the laws of thought as such. On the contrary: just as we can take what symbols we like for the objects of which we are thinking, so can we devise rules for their transformation at will. No set of rules (including those we follow in everyday discourse) has any absolute validity, nor is any involved, as earlier logicians thought, in the nature of reason. Rather we must say that every logical law owes its 'truth' to our determination to follow it, and holds good only so long as that determination is maintained.

It is clear enough that, if this view of logic is correct, the suggestion that logic throws light on the nature of mind is a false one. Now it certainly is the case that modern logicians recognize far more principles as involved in formal reasoning than did their predecessors in the science, and that they have succeeded in grouping these principles in a plurality of logical systems. But as we saw in discussing this point in an earlier chapter (pp. 95 ff),[24] it is by no means obvious that these different systems are genuine alternatives to each other. When we compare one system of formal logic with another, or when we say that inside any given system there must be no inconsistencies or that the consequences deduced must *follow* from the original premisses, we appear to be appealing to a more fundamental logic which they all presuppose. Such a logic might well be taken as concerned to define what is meant by logically possible, and this subject at least is not unconnected with the nature of our minds. For, as Kant pointed out, only minds of a certain sort would draw the distinction between the possible and the actual that we draw. Leibniz said that the truths of logic were true for all possible worlds, but he should have added 'as conceived by *discursive* consciousness'. The notion of logical possibility, which logic is concerned to elucidate, is one which to a different sort of understanding from our own would be entirely without meaning. An intuitive intelligence would neither need nor show interest in any system of logic.

It appears from this that the form of empiricism we are considering has not succeeded in making out its contention that the study of logic throws no light on the nature of mind. But there is a further difficulty in which it is involved. We have seen that its official view of philosophical propositions is that they are all analytic. But though this account can be applied with fair plausibility to a good many philosophical statements, it is not clear how it would apply to the fundamental principles of theory of knowledge. If we say, as supporters of this view do, that every significant (synthetic) statement must be, in principle at least, verifiable in some sort of sense-experience, is that to be taken as an analytic or a synthetic proposition? Certainly if someone asked why he should accept such a criterion of significant statements, it would not be enough to reply that it was a matter of definition and nothing more; he would want to be shown that no other criterion was, in the nature of things, appropriate or even possible. But a demonstration of that sort could only proceed by pointing to the *facts* about human knowing as revealed in philosophical reflection, i.e. by taking account of the constitution of our intelligence. Is this not to admit that theory of knowledge is a study which both issues in and depends upon our knowing certain *synthetic* propositions, and if so how are empiricists to explain that knowledge?

Probably the best line for them to take in this difficulty would be to admit that epistemology does involve factual knowledge of the constitution of the human intelligence, but to say that such knowledge is psychological. This would preserve their main contention, that philosophy can contribute nothing to the study of mind, at the cost of transferring part of it to another discipline. And it would be in accordance with the tradition of empiricism as represented, for instance, by Hume, whose philosophy of human nature was quite openly a psychological study.[25] But there are, of course, objections to any such procedure. In the first place (though this is far from decisive) one would expect, if epistemological questions did fall within the province of psychology, that their solution would present less difficulties than it in fact does. Like other empirical inquirers, the psychologist has a means of checking his theories, by referring to the facts they are alleged to explain. But if philosophical theories depend on facts, they are not facts of the same easy and obvious kind. And there is a second and more serious difficulty. Psychology is a first-order study, concerned to reflect on the phenomena of mental life and discover the laws they embody. But theory of knowledge is a second-order discipline, which comes into existence only when first-order studies like psychology already exist. This seems to imply that the facts which theory of knowledge elicits are on a different level from those established by the psychologist. Because theory of knowledge is concerned, not with knowing about any particular department of reality, but with knowing as such, there is a sense in which the truths it formulates are prior to those discovered in other branches of learning. It is therefore paradoxical to lump theory of knowledge in with psychology, which is, after all, a special science like any other.

These considerations suggest that the attempt of the modern empiricist school to dispose of our problem by denying its existence is not likely to be successful. In theory of knowledge, at least, there is a sense in which we are exploring the nature of the human intelligence, and so attaining a species of self-knowledge. But we are still without any satisfactory account of the relationship of this knowledge to that provided by the psychologist, our only clue being the statement that theory of knowledge is a second-order and psychology a first-order study. Let us therefore turn to a different view of the matter.

(II) We might expect an intelligent rationalist, confronted with the actual development of a science of psychology, yet anxious to maintain that philosophy throws light on the self, to argue on the following lines.[26] There can be philosophical knowledge of the self, and there can be psychological knowledge of it. When we investigate ourselves in philosophy it is the self *qua* rational, the activities of reason and understanding, that we make our object. This is the part of the self which is often called the mind, and if we adopt this designation we must say that philosophy alone can give us organized knowledge of mind. And an indispensable condition of our being able to acquire such knowledge is that we should have a faculty of rational reflection, a power of turning our attention to our own thinking and considering questions about its formal nature and presuppositions. Rational reflection here is a form of self-awareness which is *sui generis*; it is quite different from the introspection of the psychologist. Introspection is akin to sense-perception, and in it we can distinguish two factors: the apprehending of a content and the characterizing of it in judgement. Rational reflection appears, by way of contrast, to be single rather than double, intuitive rather than discursive.

But, of course, the activities of the self are not all rational, and indeed it might be true to say that singularly few of them are. Rational thinking (and rational willing, too, if there is any such thing) is conditioned throughout by the presence of non-rational factors in the self, factors which are responsible for the constant discrepancy between thinking as it is and thinking as it should be. And this leaves room for a further study of the self over and above that of the philosopher. If it is the business of philosophy to consider the self *qua* rational, it is for the psychologist to investigate the non-rational background to our rational activities – to give an account of what Professor Collingwood called 'the blind forces and activities in us which are part of human life as it consciously experiences itself, but are not parts of the historical [rational] process: sensation as distinct from thought, feelings as distinct from conceptions, appetite as distinct from will'.[27] A study of these blind forces is clearly indispensable if we are to say anything about ourselves in the concrete, since they make up a very real and important part of human nature. But just as it would be wrong to neglect them, so would it be improper to equate them with the self as such. Whatever the importance of psychology, it cannot be held to be the sole science of the self. Psychology can tell us nothing about the higher forms of human activity, just as it cannot explain how we ought to act and think if we are to be rational beings. It is to philosophy that these tasks belong.

What this comes to, put very broadly, is that philosophy studies the higher self, psychology the lower. Now it may be objected that this sharp division does not square with the facts, and would be disputed by the philosopher and the psychologist alike. It is not true, in the first place, that the philosopher is silent about the non-rational part of the self. Any theory of knowledge must clearly have a good deal to say about sensation, just as any theory of morals must discuss the nature of appetite. Nor again would the psychologist agree to leave what we are calling the higher activities of mind out of account – to abandon all consideration of the phenomena of religion and the moral life, for instance. He would argue that the influence of the factors in which he was interested could not be confined to any particular sphere, but might be seen in any department of experience, and would accordingly claim that there was no branch of human activity which his study might not illuminate.

The objection is a perfectly just one; yet it is perhaps possible to meet it without making any substantial change in the theory. Instead of saying that philosophy and psychology study different *parts* of the self, we should say that they study different *aspects* of it. The self is to be regarded as containing two sets of factors, rational and irrational, and we can direct our attention to whichever we choose. The influence of both sets can be seen throughout our mental life: we are constantly thinking and acting as rational beings, yet we are also constantly failing, through the operation of non-rational factors in ourselves, to live up to the standards we set ourselves. To give a complete account of the self we must hence undertake both a philosophical and a psychological study of it. We cannot fuse the two, though it has often been proposed that we should, since the self *qua* rational differs fundamentally from the self *qua* non-rational. To consider myself as a thinker, or again as a moral agent, I must make use of methods totally different from those I should adopt in investigating myself psychologically. And it seems hard to deny that the difference is connected with the fact that I have a kind of insight into my rational nature which I do not have into the phenomena of the non-rational self.

§6. This theory of the relationship between psychology and what may be called, very generally, philosophy of mind, presents an account which, as it seems to the present writer, is not only intelligible but also substantially correct. It has the solid merit of doing justice to all that is important in the rationalist tradition, whilst at the same time retaining some central features of the empiricist view; and for that reason might be expected to commend itself to philosophical moderates of all kinds. But it cannot be denied that, as here expounded, the view contains certain obscurities, and that in particular the notion of rational reflection which it involves stands in need of further clarification. I will end this chapter with some remarks on that notion and on the consequences its acceptance entails for theory of knowledge.

In the first place, it seems important to point out a sense in which every philosopher ought to agree that rational reflection is possible. To see that sense we have only to take some ostensible example of thinking or rational action and (in a broad sense of the term) reflect upon it. If we do that, we find that there are two quite different questions we can ask about it. First, we can inquire whether it is in fact a genuine act of thinking or an action in the true sense. What troubles us here is the possibility that we might not be thinking or acting at all, but only behaving in an automatic way, as of course we often do. Our way of answering the question is to examine our behaviour, as revealed in introspection, closely, and apply to it such tests as the psychologist may devise. But once we have convinced ourselves that we are really thinking or acting, there is a further question we can go on to ask: what is the content of our thought, or the purpose to which we are striving to give effect? It is here, I think, that rational reflection comes in. Reflection on our rational acts enables us to grasp the content of those acts: to find out what we are thinking, or what we are trying to do. It is true that reflection on any one act will not answer all questions about it: it will not enable us to connect it with previous acts, to get at the premises or motives which lay behind our thought or actions. To discover those an extensive inquiry may be needed. But that we can grasp what it is we are thinking when we think, and do so by rational reflection, seems obvious enough.

However, to say that rational reflection is possible in this sense is not to say anything exciting. It amounts to no more than the assertion that human beings have a faculty of abstraction, and can consider the content of their thinking in separation from its physical background. The theory we are considering would certainly claim that this was true, but it would also claim something more. It would claim that rational reflection gives us insight of a special kind into the nature of reason. Thanks to this insight the philosopher is able to discover certain truths – indeed, certain necessary truths – about the human mind, and so to provide an important form of self-knowledge. It is on the possibility of rational reflection in this wide sense that the whole theory turns. What are we to say about it?

Here I think we should do well to take the cases of logic and theory of knowledge separately. Let us take first the logician reflecting on his own acts of thinking and seeking to establish the formal principles to which they conform. Aristotle said long ago that we could be led by rational reflection to see that our thought embodies certain formal principles such as the law of contradiction. His own explanation was that the law is a necessary truth about the world of fact, discoverable by intuitive induction. In criticizing this view in an earlier chapter[28] we pointed out that the true concern of formal logic is not with the actual but with the possible, and that the laws it establishes are prescriptive rather than factual in character. But it may be asked here whether we should not say that they are both prescriptive and descriptive: prescriptive to whatever falls within experience, but descriptive of the essential nature of the experiencing subject. To this we can certainly answer that they are not descriptive of mind in the way the old rational psychology purported to be, since if anything is clear it is that the logician has no direct intuition of his own thinking self. Yet it seems perverse to deny that, in knowing that the mind, if its thinking is to be effective, must conform to certain principles, we are discovering something important about it. The logician, though unable to study the subject-self directly, is none the less able to bring out some aspects of it quite clearly. What he does can be compared to what, on a Kantian view of the science, the Euclidean geometer does about space. According to such a view, there is an essential connexion between geometry and space, yet space itself cannot be intuited.[29] We find, however, that when we construct figures in space, or consider the geometrical properties of objects in space, we can discover certain *a priori* principles to which all such figures and objects conform. These principles, if Kant is right, depend on the necessary nature of space itself, and in knowing them we are thus gaining insight into that necessary nature. Similarly with logic: unless we adopt the modern empiricist account of them, it seems that we must agree that logical principles depend upon the nature of the knowing subject, and that in establishing them the logician gains insight into that nature, knowing it not directly but obliquely.[30]

But if we accept this conclusion, as I think we must, it is only right that we should be clear that it does involve a most important concession to rationalism. We are allowing that, in logic at least, mind can know its own nature, and can do this by what is, in fact, a species of intellectual intuition, though a rather odd one. It is true that all that emerges from a logical study, even on this account, is knowledge of the most general principles by which we are bound in our thinking. If reason knows itself in logic, such knowledge is, as Kant was never tired of saying, of a purely formal character. But whether formal or not, it is knowledge which none the less deserves its name, and for that reason excites the interest of the philosopher.

Now let us turn to theory of knowledge and consider whether we can give the same account of that discipline. At first sight it appears that we can. If the task of the logician is to discover the principles which govern valid thinking as such, the epistemologist may be held to consider thinking of a special type – namely, *a priori* thinking of objects – with the aim of eliciting its special principles. But when we examine the way in which this task is carried out, we cannot avoid seeing an important difference between logicians and epistemologists. The difference is that while the logician seems to have intuitive insight into the principles of formal reasoning, the epistemologist must have recourse to an elaborate proof to establish his conclusions. We cannot just *see* that the law of causality is involved in objective thinking as we can see that the law of contradiction is involved in formal thinking.

And there is a further difference. The business of theory of knowledge does not end with the establishment of the *a priori* principles governing objective thinking: it has also to consider in virtue of what we need to apply those principles, and this is indeed its central problem. Here reason does truly seek to elicit its own nature and bring out the properties of the subject of knowledge. But though the procedure of reason in solving this problem is without doubt to reflect on examples of knowledge and consider what is involved in them, it would be absurd to claim that the result of the process is a flash of intuitive insight. What does result is a *theory* which seeks to do justice to the facts of consciousness and must be tested by reference to them. It is the facts which are known by rational reflection rather than the theory about them. And just how much argument may be needed to establish any set of conclusions we can see by considering any of the classical writings on the subject: Hume's *Treatise*, Kant's *Critique* or Hegel's *Logic*.

I conclude that, though in theory of knowledge our aim is to throw light on the knowing subject, we cannot claim any special intuition to help us in the search or to guarantee our results. Our procedure, here as in the positive sciences, is to review particular cases in the hope of finding a theory which will fit them all. The more complex our facts, the more argument we need to ground our conclusions, and the less certain we can be of them. But if I am asked whether this does not show that theory of knowledge should be considered a positive science like any other, I can answer only by pointing to the difference between first-order and second-order inquiries, and by saying that there is a sense in which epistemology presupposes the positive sciences rather than is co-ordinate with them. It seems to me, indeed, that theory of knowledge occupies a peculiarly ambiguous position in the world of learning: that it is neither an ordinary empirical discipline, because it is concerned with the presuppositions of experience itself, nor yet an *a priori* one, since there is no insight to which we can appeal for a definitive solution of its problems. What status its propositions have, and how they can be tested, seem accordingly to need careful discussion. But I have not found any such discussion in the main authorities I have consulted, nor can I supply one myself.[31]

[1]Of *Reason and Experience*.

[2]There is a sense in which *all* sense-data are private to a single percipient, since every act of sensation is an act of a self. But in the case of the external senses we find that one man's data correspond to, or cohere with, another's; and when this happens we say that they are observing the same object. There is no corresponding phenomenon in internal sensation: even when my feelings are sympathetically like yours. I do not suppose that we are both aware of the same self.

[3]On p. 21 of *Reason and Experience*. The quotation was from Hume's *Treatise*, ed. Selby-Bigge (Oxford: Clarendon Press, 1888), p. 252.

[4]See A. J. Ayer, *Language, Truth and Logic* (London: Gollancz, 2nd ed. 1946), ch. ii, for an excellent exposition of the modern empiricist view of the function of philosophy.

[5]*Principles*, § 27 (first edition text).

[6]Ibid., § 142 (second edition text). (Berkeley's account of spirit, so far as it goes, is to be found in §§ 27 and 135–44.)

[7]*Anthropologie in pragmatischer Hinsicht*, footnote to § 4 (*Kants gesammelte Schriften*, Prussian Academy edition, vol. VII, 134 n.). These lectures were published by Kant himself in 1798. The quotation given is hardly a fair specimen of the difficulty of this 'popular' course. (The *Anthropologie* is translated by Mary J. Gregor as *Anthropology from a Practical Point of View* [The Hague: Nijhoff, 1974]. The reference is to p. 15 n.)

[8]*Kants gesammelte Schriften*, vol. IV, 471.

[9]Kant must in any case be given the credit of having clearly seen that empirical psychology is a branch of natural knowledge and not a part of metaphysics. For his views on this subject see *Critique of Pure Reason*, A 848–9/B 876–7.

[10]*Critique of Pure Reason*, B 157.

[11]*Critique of Pure Reason*, B 422–3 n.

[12]Ibid.

[13]Of *Reason and Experience*.

[14]e.g., A 382.

[15]Ayer, p. 126. Mr Ayer refers his readers to Mr. Russell's *Analysis of Mind*, ch. ix, for an elaboration of the point. For arguments against an account of cognition in terms of subject and object see ch. i of that work.

[16] T. D. Weldon (*An Introduction to Kant's* Critique of Pure Reason (Oxford: Clarendon Press, 1945), pp. 154 ff.) has even argued that Kant could find no content for inner sense except such past acts of awareness.

[17]i.e., in *Reason and Experience*.

[18]Hegel: *Encyclopaedia*, § 34 (*Zusatz*), translated by W. Wallace as *The Logic of Hegel* (Oxford: Clarendon Press, 2nd ed. 1892).

[19]Compare, e.g., *Critique of Pure Reason*, A 401–2.

[20]*Grundlegung zur Metaphysik der Sitten*, section ii, last paragraph. (The *Grundelgung* is translated by H. J. Paton as *Groundwork of the Metaphysic of Morals* [New York: Harper, 1964].)

[21]For Kant's views on philosophy see especially *Critique of Pure Reason*, A 832 ff./B 860 ff.; *Logik* (ed Jäsche), Introduction, i, iii; *Grundlegung zur Metaphysik der Sitten*, Preface. (The *Logik* is translated by R. Hartman and W. Schwartz as *Logic* [Indianapolis and New York: Bobbs-Merrill, 1974]. For translation of the *Grundlegung*, see previous footnote.)

[22] *Kants gesammelte Schriften*, vol. IX, 14; p. 16 of the translation by Hartman and Schwarz. Cf. *Critique of Pure Reason*, B ix.

[23] What *is* a proposition? Here we get divergent answers from different exponents of formal logic. One group, now rather outmoded, thinks of propositions as in some sense objective constituents of the universe, constituents which exist whether or not anyone takes up a mental attitude to them. Seeing that one proposition entails another, for this group, is to discover something important about the nature of fact. But the empiricist school whose theories are discussed in the text is far more inclined to identify propositions with sentences rather than facts, and to look on formal argument as a means of transforming one sentence or set of sentences into another. Logical argument is, in this interpretation, purely tautological.

[24] Of *Reason and Experience*.

[25] Compare the way in which Hume establishes, in the first section of the *Treatise*, his own version of the main principle of empiricism, that every simple idea must be derived from some precedent impression.

[26] In what follows I have had in mind particularly the views of Professor R. G. Collingwood, though I have not set them out in detail, as I do not wish to discuss here his contention that all philosophical knowledge is historical. For Collingwood's opinion of the province of psychology see especially *The Principles of Art* (Oxford: Clarendon Press, 1938), p. 171 n. and *The Idea of History* (Oxford: Clarendon Press, 1946), pp. 230–1.

[27] Ibid., p. 231.

[28] Of *Reason and Experience*.

[29] Kant himself frequently says that space *can* be intuited; but it is not clear that he ought to do so, for the purposes of his philosophy of mathematics at least.

[30] I am, of course, disregarding here the third possibility that logical laws might be laws of things and not of thought: on this see *Reason and Experience*, p. 98. On my own view, logic must apply to whatever exists, since the minimum prerequisite of what we take to be fact is that we should be able to think it.

[31] Perhaps I should mention that Kant was led to overlook the need for such a discussion by considering critical philosophy a part of metaphysics, and so making the answer to the question 'How is metaphysics possible as a science?' also answer 'How is critical philosophy possible?' But to show that metaphysics can consist of valid synthetic *a priori* judgements is not to show that critical philosophy consists of propositions of the same kind.

IX

THE AGE AND SIZE OF THE WORLD

JONATHAN BENNETT

KANT rejected two views about the world: that it is infinitely old and that it is infinitely large. But he failed to make himself clear. One cannot be sure what his point is about the infinite age and infinite size of the world, and I haven't found the commentators very helpful either (see Section II below). In this paper I present a thesis about what was really troubling Kant in regard to those infinities, and about what solution he proposed for his troubles.

I. THE AGE ARGUMENT

Kant thinks of the world's past as a series, and equates the world's being infinitely old with this series' having infinitely many members. I shall speak of the series of past *events,* using 'event' as a purely technical term to mean 'one minute's worth of world-history'. It could as well be a year's worth or a century's worth, just so long as it isn't construed as anything like: whatever happened in the past hour, whatever happened in the half-hour before that, whatever happened in the quarter-hour before that, and so on; for that series can have infinitely many members without taking us back as far as lunch-time. Also, of course, the members of the series of past events must not be allowed to overlap one another.

When Strawson discusses this matter he pretends that 'for as long as the world has existed, a clock has been ticking at regular intervals', and he then equates the world's age with the length of the series of past ticks.[1] His 'ticks' do exactly the same work as my 'events'.

Now, Kant argues like this. If the world never began, then it has been going on for ever, and the series of past events – past ticks of Strawson's clock – has infinitely many members. But:

The infinity of a series consists in the fact that it can never be completed through successive synthesis. It thus follows that it is impossible for an infinite world-series to have passed away. (A 426/B 454.)

From *Synthese* **23** (1971), pp. 127–46. Copyright © 1971 by D. Reidel Publishing Company, Dordrecht-Holland. Reprinted by permission of D. Reidel Publishing Company.

That *is* Kant's argument – his presentation of the alleged conceptual difficulty in the idea that the world is infinitely old. The argument looks bad, because on the face of it it is open to an obvious objection. Kant says that 'the infinity of a series consists in the fact that it can never be completed through successive synthesis' – that is, through a one-by-one enumeration of its members – but that is just false. A series of the sort Kant has in mind must, if it is infinite, be open at one end; it cannot have both a first and a last member; and so the enumeration of its members, *if started*, 'can never be completed'. But such an enumeration could be completed all the same, if it did not ever start but had been going on for ever.

Let '*T*' name a known point in past time, say the moment when you began reading this paper. Then Kant's argument can be put thus: the series of events-before-*T* was completed at *T*; events could in principle be counted as they occur; and so a counting or enumeration of the series of events-before-*T* could have been completed at *T*; *and so* the series of events-before-*T* does not have infinitely many members. But – the obvious objection runs – that final step is not valid, or anyway Kant hasn't shown that it is. For he hasn't displayed any incoherence in the idea that at *T* someone said '*T* minus 0', and a minute earlier said '*T* minus 1', and so on – through *every* event-before-*T*, there being no earliest such event.

II. COMMENTATORS ON THE AGE ARGUMENT

Of the commentators who discuss this matter in books on Kant, the earliest I have read is Caird, who seems content with Kant's argument and wholly unaware of the obvious objection to it.[2] Kemp Smith rejects Kant's conclusion indignantly, but hasn't the patience to look carefully at the argument Kant uses.[3] According to Kemp Smith, apparently, Kant's premiss is that 'we cannot comprehend how, from an infinitude that has no beginning, the present should ever have been reached', which I find un-Kantian and unintelligible. Kemp Smith seems to find it *true*, but says that it does not justify us in 'denying what by the very nature of time we are compelled to accept', namely that 'time is . . . infinite, alike in its past and in its future'. All that is unhelpful because sheerly irrelevant, which one can't often say about Kemp Smith. In Section IV I shall introduce another of his remarks which is not irrelevant but deeply and precisely wrong.

Ewing does at least expound the obvious objection to Kant's argument. He says that it accuses Kant of a 'puerile fallacy', which seems to me a bit strong; and he then proceeds to defend Kant against the obvious objection; but the defence seems to be quite incoherent.[4] Weldon's treatment of

Kant's argument is rather cursory and, in my opinion, not nearly critical enough.[5] Gottfried Martin's anxiety to see Kant's argument as an implied commentary on earlier philosophers leads him to misrepresent it to an extent that must be seen to be believed.[6]

Other writers, such as Benardete, expound Kant's argument faithfully and attack it with the obvious objection.[7] Strawson also expounds the obvious objection and seems to regard it as fatal:

A temporal process both completed and infinite in duration appears to be impossible only on the assumption that it has a beginning. If . . . it is urged that we cannot conceive of a process of *surveying* which does not have a beginning, then we must inquire with what relevance and by what right the notion of surveying is introduced into the discussion at all.[8]

I wholly agree with this. But, apparently unlike Strawson, I am sure that the notion of surveying *has* a right to be introduced into the discussion, and indeed given a crucial place in it. I shall defend this later.

III. THE SIZE ARGUMENT

If the world is infinitely large, Kant thinks, then the thought of *the size of the world* must be the thought of *every member of a series of finite world-parts* – e.g. a series of non-overlapping cubic miles of world. The size of something finite can be regarded as what Kant calls 'the magnitude of a quantum which is . . . given in intuition as within certain limits': one need not think of its size serially, because, being finite, it is the sort of thing that might in principle be perceived in its entirety all at once. But if a thing's size is infinite, then:

[its] magnitude can be thought only through the synthesis of its parts, and the totality of such a quantum only through a synthesis that is brought to completion through repeated addition of unit to unit. (A 428/B 456.)

The point is spelled out, perhaps helpfully, in a footnote:

The concept of totality is in this case simply the [concept] of the completed synthesis of its parts; for, since we cannot obtain the concept from the [perception] of the whole – that being in this case impossible – we can apprehend it only through the synthesis of the parts viewed as carried, at least in idea, to the completion of the infinite.

It is easy to guess how the argument will run from that point. Kant will object to the idea of an infinitely large world for the same reason that he objects to the idea of an infinitely old world:

In order, therefore, to think, as a whole, the world which fills all spaces, the successive synthesis of the parts of an infinite world must be viewed as completed, that is, an infinite time must be viewed as having elapsed in the enumeration of all co-existing things. This, however, is impossible.

This is a most peculiar argument. Does Kant assume that if the world is infinitely large then the series of past events is infinite? If his argument depends on that, it surely fails. Perhaps he is assuming only that if the world is infinitely large then the series of past events *could be* infinite, and arguing from this that since the series of past events cannot be infinite the world is not infinitely large. But it is not at all clear how Kant proposes to justify the initial assumption. What, for example, can we make of the following way of putting the point? –

Unlike time, space does not in itself constitute a series. Nevertheless the synthesis of the manifold parts of space, by means of which we apprehend space, is successive, taking place in time and containing a series. (A 412/B 439.)

Granted, a region of space *can be thought of* serially, e.g. as some small region, plus a yard-thick shell around it, plus a yard-thick shell around that, and so on. Granted also, a large enough region of space *must be apprehended* serially, so that the actual exploration of it would 'take place in time and contain a series'. But how do we get from those two concessions to Kant's view – if it *is* his view – that if an infinite series of operations cannot be completed then the world is not infinite in extent?

I think that Kant entirely fails in his attempt to present the difficulty about the world's size as a special case or upshot or corollary of the difficulty about the world's age. I shall later argue that the attempt should never have been made – that the problem which Kant does have about the world's size ought to have been allowed to stand on its own feet.

IV. THE SCOPE OF KANT'S PROBLEM

Although Kant denies that the world can be infinitely old or large, he thinks that it cannot be finitely old or large either. (The mistakes which his anti-finitism involves, e.g. the assumption that something of finite size must have a boundary, lie beyond the scope of this paper.) So in the area I am discussing he sees himself as having not merely two views but two *problems*, each expressible in the form: 'We have grounds for wanting to describe x as infinite, but there is a difficulty about using the concept of infinity in this way.'

We know that Kant thinks he has two such problems – two values of x –

though he hasn't made clear why they are problems, i.e. what the difficulty is about applying the concept of infinity to the series of past events or to the series of cubic miles (say) of world. All we have is an obscure reduction of the size problem to the age problem, together with an obviously defective account of the latter. One might conclude that Kant has shown unwittingly but all too clearly that his age and size 'problems' are bogus. I think that would be wrong, though; and as a preliminary to showing that it would be wrong I want to consider the question – what other problems of this general form does Kant think that he has?

I have contended that Kant ought to have allowed the size problem to stand on its own feet, rather than trying to reduce it to the age problem. But I do not mean that each should be presented just as the problem of how the notion of infinity can be brought to bear on the empirical world. Weldon sees Kant in that light, saying that according to Kant

the understanding can frame no concept of an infinite series of places or events as an actual empirical object, since nothing of this nature can possibly be given in experience,[9]

as though Kant's objection were to empirical infinities as such. But that misrepresents him, for he distinguishes clearly – or at least loudly – between infinities which do and ones which don't involve a conceptual difficulty. As evidence for this, and against Weldon, consider the following remark of Kant's:

Since the future is not the condition of our attaining to the present, it is a matter of entire indifference, in our comprehension of the latter, how we may think of future time, whether as coming to an end or as flowing on to infinity. (A 410/B 437.)

There are two points here. One is that we don't have to raise the question of whether the series of future events is infinite, whereas Kant thinks that we are forced to speculate about the world's age and size. But the quoted passage also implies clearly enough, I think, that in Kant's view we *can* suppose that the series of future events is infinite without thereby encountering any conceptual obstacle. So Kemp Smith is wholly wrong when he says:

Kant limits his problem to the past infinitude of time. The reason for this lies, of course, in the fact that he is concerned with the problem of creation. The limitation is, however, misleading.[10]

This implies that the trouble Kant finds in the infinity of the series of past events is equally present – and perhaps even that Kant knows that it is equally present – in the infinitude of the series of future events. This, I contend, is a damaging mistake.

But that doesn't explain what the line is between the infinities which Kant finds troublesome and the ones he doesn't. All we know so far is that a past infinity is troublesome while a future one isn't. What Kant says is that troublesome infinities are precisely those that lie in the past or involve the thought of an infinity that lies in the past. He expresses this by saying that an infinite series is troublesome if and only if it is a series of *conditions*, which he also calls a *regressive* series; and he says that the source of the difficulty is 'the . . . idea of the absolute totality of the series of conditions of any given [thing which is] conditioned', and he says explicitly that this idea 'refers only to all *past* time' (A 412/B 439). So the idea of an infinitely large world, though ostensibly involving a series lying wholly in the present, can be represented as a source of difficulty only by being shown to involve, covertly, the thought of an infinite series of past operations. And we have seen how unconvincing is Kant's attempt to make this move.

He would have been spared the need to make the attempt if, instead of (a) equating the troublesome/innocent line with the past/non-past line, he had (b) equated the troublesome/innocent line with the non-future/future line. For then he could treat a present infinity, such as the infinite size of the world now, as troublesome not because it covertly stretches into the past but just because it doesn't lie wholly in the future. I don't think that (b) would clash with anything solid in Kant's discussion of these matters. It would conflict with some of his remarks about 'series of conditions', and about the related distinction between 'regressive' and 'progressive' series; but these Kantian technicalities are not handled so firmly and cogently that we are forced to abide by them. Nor does (b) conflict with any of Kant's examples; for his only example of a 'progressive' or untroublesome infinite series is, precisely, that of the infinite series of future times or future events.

So we can fairly safely pretend that Kant's basis is (b) rather than (a). I now proceed to argue that this pretence brings positive advantages.

V. STARTING INFINITE TASKS

Kant's approach to any empirical concept is dominated by his view that anything I can intelligibly say about the empirical world must be interpreted somehow in terms of what I could, in principle, discover for myself by my own observations. This is a sort of first-person phenomenalism which is embodied in, among other things, Kant's theory that our concepts are just tools for the orderly management of our sense-impressions. This raises a question about the concept of infinity: how can I have any legitimate use for that concept in application to the empirical world? what experience of mine could possibly require me to make any use of it? can I even intelligibly suppose myself to have experiences which justified a use of it? Kant raises these doubts by suggesting that the past of an infinitely old world, like the size of an infinitely large one, is 'too great for the understanding' – i.e. so great that we can't have a concept of it. This starts to sound like Weldon and Kemp Smith, but I add one qualification which they omit – namely, that the concept of an infinite future is not, even by Kant's phenomenalistic standards, 'too great for the understanding'.

The point I am making has been interestingly developed in Dretske's paper 'Counting to Infinity'.[11] Dretske contends that it is possible – or at any rate only medically impossible – that someone should count all the natural numbers. He argues that we can intelligibly suppose that someone counts to 100, say; and if we can intelligibly suppose that someone counts to n then we can intelligibly suppose that someone counts to $(n + 1)$; and so it makes sense to suppose that someone has just begun to count, and is going to count every natural number. There will of course never be a time at which he has counted them all, but given any natural number a time will come when he will have counted *it*.

Dretske's conclusion seems to me absolutely right. If it chokes you, dilute it a little: say of our supposed counter not that he will count all the natural numbers but that he will count each natural number. The basic point is just that we can make sense of the idea of beginning on some task and never stopping. Similarly, we can make sense of the idea that we shall last for ever: many people believe that they will last for ever, and I can see no incoherence in this belief, merely falsity.

Since many people don't see why Dretske is right, I shall linger for a paragraph. The statement that *I shall count all the natural numbers* is expressed by

$$(n) \, (\exists t) \, (n \in N \rightarrow \text{I count to } n \text{ before } t) \qquad \text{(A)}$$

where t ranges over times and N is the set of natural numbers. Those who protest, against Dretske, that I couldn't ever *complete* the counting of all the natural numbers are implying that *I shall count all the natural numbers* is equivalent to

$$(\exists t) \, (n) \, (n \in N \rightarrow \text{I count to } n \text{ before } t). \qquad \text{(B)}$$

It does seem natural to think that if I shall count them all I shall eventually *have counted* them all; or to think that if it is true of each of them that I shall eventually have counted it, then I shall eventually have counted all of them.

Let us try to express this natural assumption in quantificational terms. The difference between (A) and (B), as expressed by the order of the quantifiers, is that between a weaker and a stronger statement – like the difference between *Everyone has a friend* and *someone is everybody's friend*. So we cannot derive (B) from (A) without adding further premises about counting or priority or numbers or the like. The basic relevant fact about counting is that if I count to n before t then I count to every lower number before t, which is to say that

$$(n)\,(t)\,((n \in N \,\&\, \text{I count to } n \text{ before } t) \rightarrow (m)\,(m < n \rightarrow \text{I count to } m \text{ before } t)). \tag{C}$$

But (A) and (C) together still don't yield (B). An addition which does permit the derivation of (B), and apparently the weakest one that will do the job in a manner relevant to our present theme, is

$$(\exists n)\,(n \in N \,\&\, (m)\,(m \in N \rightarrow m \leqslant n)) \tag{D}$$

which says that there is a highest natural number, i.e. that the set of natural numbers is finite. The derivation of (B) from (A), (C), (D) depends upon no extra assumptions about counting etc.: it goes through, quite formally, with 'count to . . . before . . .' replaced by an arbitrary two-place predicate. I conclude that those who say that I shan't count all the natural numbers because I shan't ever have counted them all are ignoring a distinction – namely that between (A) and (B) – which is usually negligible but which is important in just such contexts as Dretske's, where we don't have (D) because what is being counted is an infinite set.

VI. Completing Infinite Tasks

So much for the statement that I shall perform an infinite task. What about the statement that I have performed an infinite task? Dretske says 'I'm not sure that this makes sense'. His doubts do not concern the abstract logic of the statement, which, he shows, mirrors the logic of the statement he finds untroublesome. In particular, just as someone who will count to infinity won't ever have finished, so someone who has counted from infinity wasn't ever not-yet-started. 'And *that* is why the supposition doesn't make sense: for a task which one doesn't ever start is a task on which one isn't ever engaged and which one can therefore never finish.' That is wrong. It assumes a principle which is valid for finite tasks but not for infinite ones –

the logical points involved being exactly those displayed in the preceeding paragraph.

Yet Dretske, like Kant, doubts the intelligibility of the supposition that one has completed an infinite task. The source of this doubt presumably involves facts about what it is to be a person, or to perform a task, or live through an event, or the like. Wittgenstein, for some purpose, once invited his hearers to image coming upon a man saying '. . . nine, five, one, four, one, three, phew!' and then announcing that he had just completed a backward recital of the entire decimal expansion of π.[12] The conversation might go on like this: 'All of it?' 'All of it'. When did you begin?' 'I didn't begin, of course. I have always been reciting the decimal expansion of π, until just a moment ago when I finished – thank God!' If someone claimed to be embarking on a forwards recital of π, we shouldn't believe him, but we could understand what he said: we can take in the idea of doing so much, then a bit more, and, however much he had done, *always* a bit more still. But the creepiness of Wittgenstein's story, like Dretske's hesitancy, suggests that there is a conceptual difficulty in the idea of someone's completing an infinite task upon which he has always been engaged. It isn't clear that this is intelligible to us as a possible state for a sentient being.

The view that it is *not* intelligible – 'the Kantian view', for short – is fairly widespread, and I am inclined to accept it. If it contains any truth, I think it must be for reasons of the following sort: – The notion of someone's having performed a series of operations – if 'some*one*' is taken seriously – involves the notion of his remembering performing those operations, or knowing what it is like to have performed them, or in some way *possessing* that part of his past. How much I have done or undergone is a kind of measure of how much of me there is now. And so, to suppose that I had performed an infinite series of operations is to suppose myself to be, *now*, infinitely experienced, or endowed with an infinite stock of memories, or something of that kind. And it can plausibly be maintained that that cannot be supposed. In contrast with this, the supposition that I shall perform all of an infinite series of operations does not involve the idea of my possessing, now or at any future time, anything like an infinite stock of memories. This contrast arises from a fundamental asymmetry in sentient beings: they have more epistemic grasp of the past than of the future. I think this is a necessary truth. If it is not, then the possible sentients who falsify it won't accept the Kantian view and won't see any force in Kant's discussion of the age and size of the world.

A backwards recital of the natural numbers or of the decimal expansion of π is mechanically generated by a rule, and so it arguably burdens the memory with nothing more than the rule. But that feature of Dretske's and

Wittgenstein's examples is just an expository convenience. Our concern is with the Kantian notion of a sentient being's epistemic grip on the contingencies of his past experience; and if we aren't to drift away from that, and thus from what is philosophically interesting in this area, we must now think in terms of non-rule-generated tasks or biographies, in which each episode is a partly brute-fact addition, imposing at least some extra load on the memory. For the same reason, we can ignore the boring possibility that someone should have lived for ever but at no time have memories stretching further back than, say, 100 years.

So much for infinite past versus infinite future. As for infinite present: there is clearly no room for that notion while we are concerned with what one can envisage oneself as encountering in experience. For example, there can be no question of supposing oneself to know the world to be infinitely large because one perceived it, all at once, in all its infinite extent. In any Kantian spelling-out of things in first-person phenomenalistic terms, each *present* must be extremely thin: the only way to build up a thick story – e.g. one which gives work to the concept of infinity – is by stringing together a series of presents. These will stretch into either past or future, and so they are covered by the previous discussion.

Perhaps this is what Kant is getting at in his purported reduction of the world-size problem to the world-age problem. His point there may be that the world cannot be infinitely large if 'the world's size' has to be elucidated in terms of what one would have experienced by the time one had ransacked the entire world, together with the point that that kind of elucidation seems to be implied by Kantian phenomenalism. But it must be confessed that if that is Kant's thought, then he expresses it most unclearly.

The unclarity can be explained. Kant thinks that the conflict between finitism and infinitism creates a problem which can be solved by appealing to a theory of his – 'transcendental idealism' – whose only intelligible component is precisely the phenomenalism I have been discussing; and he claims that his theory's ability to solve this problem is a powerful argument in its favour. But of course that argument is viciously circular if the theory is also required to create the problem in the first place; and I conjecture that this is one reason why Kant is less than candid, or less than clear, about phenomenalism's role in creating an objection to the thesis that the world is infinitely large. Phenomenalism is in fact also involved in creating the other side of the size-problem, i.e. in Kant's objections to the world's being only finite size; but that lies far beyond my present scope.

The reasons I have given for the Kantian view are assailable. In particular, I have no adequate answer to the following objection:

'I can suppose myself to have an infinite stock of memories, so long as I think of them as possessed dispositionally – which after all is how we do possess most of our knowledge of all kinds. My life so far has given me an accumulation of memories which are registered in me now as my ability to answer many questions about my past. The supposition that Kant thinks I cannot make is just that I should now be able to give – should now dispositionally know – the right answers to infinitely many distinct questions about my past. But of course I can suppose this. I can suppose myself able to answer five questions; and if I can suppose myself able to answer n then I can suppose myself able to answer $(n + 1)$. So the Kantian view is false – or at least your defence of it doesn't work.'

One possible reply, anticipating a line of thought which I shall exploit in Section X below, is that the objection takes my epistemic possession of an infinite past to consist in my having certain abilities, i.e. in a fact about myself in a possible *future*. Furthermore, the objection had to do this. Memories may be episodic, occurring as states of consciousness whose relation to one's past is logically similar to that between one's sensory states and one's present objective environment. What cannot be supposed is that one should at any time have infinitely many memories of *that* kind. Any infinite stock of memories must be mostly dispositional, and so the infinity it involves must be in a certain sense projected into the future.

That reply, though I think it has some force, is less than compelling. I hope one can do better for the Kantian view that I have so far succeeded in doing. But my exegetical purposes don't require me to defend the Kantian view. It suffices that it is a view which Kant was inclined to hold, even if he didn't quite bring it to the level of consciousness. Given that much, I can explain some aspects of his thought which in the literature that I have read have been left as mysteries.

VII. The Official Solution

Before showing what really goes on in Kant's problem-solving endeavours regarding non-future infinities, I must sketch what he says is going on. On the face of it his official 'solution' is no solution at all, not even a bad one, but merely an inert piece of dogmatising. It turns out, though, to be a cover for two quite different problem-solving moves. One of these is mistaken, but they both have life in them. The relation between them cannot be properly understood except on the basis of a grasp of how each shelters under the dead 'official solution', to which I now turn.

According to Kant, a way out of the impasse is opened up by the realization that the world 'does not exist in itself', a claim which can also be expressed in Kantian language by saying that the world is a *phenomenon*.

Kant's view is that a phenomenal item, but no other sort of item, can avoid being either finite or infinite, and so he is entitled to say: 'I . . . deny the existence of an infinite world, without affirming in its place a finite world.' (A 503/B 531.) 'If we regard the two propositions, that the world is infinite in magnitude and that it is finite in magnitude, as contradictory opposites, we are assuming that the world . . . is a thing in itself . . .' (A 504/B 532.) But it is not a thing in itself, and so we need not opt for either proposition.

When Kant says that the world is not a thing in itself, he means two sorts of things. (a) We are trapped on this side of the veil of perception: we cannot know 'things as they are in themselves' but only 'things as they appear to us', and so the world we know is only an assemblage of 'things as they appear'. (b) All our concepts are tools for the intellectual handling of our sensory intake: we cannot make sense of any statements about the world except ones admitting of a broadly phenomenalist analysis. Kant often has in mind both (a) and (b), regarding them as parts of a single doctrine called 'transcendental idealism'. But in fact (a) is condemned by (b): on Kant's own theory of what our concepts are, all our thinking is restricted to thoughts about actual or possible *data*, items which could be given or presented or made to appear to us; and so we cannot make sense of the notion of 'things as they are in themselves', i.e. the notion of something considered as having a certain nature which is not to be grasped or elucidated in terms of how the thing might appear to us. I shan't expand on this point, as I have already done so in my book, as has Strawson in his.

Kant too often discusses infinity in the spirit of (a) rather than (b), implying that if the world were radically 'out there', beyond the veil, it might be infinite, but that since it is only an 'appearance' it cannot be infinite even if it isn't finite either. Such remarks, taken just as they stand, seem to me to be worthless – not just false, but dead.

If we are to salvage anything from Kant's use of the notion of 'things in themselves', as it occurs in the context of his infinity problems, we shall have to stress (b) rather than (a). Roughly, we shall have to construe Kant as saying that a certain difficulty about non-future infinities can be removed by taking a phenomenalist approach to statements about the world, e.g. about its age and size. I believe that we can construe him thus. In the material that Kant presents us with, there are two strands which could be expressed in the form: 'Since statements about the world are to be understood phenomenalistically, the problem about non-future empirical infinities can be solved as follows . . .' One strand maintains that phenomenalism shows us how the world can be neither finite nor infinite, whereas the other maintains that phenomenalism shows us why it is not after all objectionable to suppose that the world is infinitely large and infinitely old.

VIII. The 'Weakening' Move

Of Kant's two purportedly problem-solving moves, I take first the one which doesn't work. I shall expound it in connection with the world's size: its re-application to the world's age is a routine matter, as will eventually become clear.

According to Kant's phenomenalism, any statement about the world is equivalent to a statement about actual and possible experiences. More specifically, any statement about how large the world is is equivalent to a statement about how long a series of experiences one could have, each consisting in the exploration of a hitherto-unexplored stretch of the world. It is very important that for Kant these are genuine *equivalences*:

Only by reference to the magnitude of the empirical regress [i.e. the series of possible explorations] am I in a position to make for myself a concept of the magnitude of the world. (A 519/B 547.)

That the series of possible explorations has such-and-such a length is not just a consequence of the world's having a certain size – it *is* the world's having that size.

So the statement that the world is not finite in size is to be analysed into the statement that the series of possible non-repetitive world-explorations has no end, i.e. that no finite series of explorations would exhaust the world, or that any finite series of explorations would leave some world unexplored. This has two different sorts of significance for Kant. The one that concerns me in the present section really has nothing to do with the notions of experience, exploration, 'empirical regress' etc. Abstracting from all such notions, we have Kant expressing 'The world is not finite in size' in the form 'No finite amount of world includes all the world there is' or 'Every finite quantity of world excludes some world'. This, I submit, seems to Kant to be a weaker statement than the statement that there is an infinite amount of world.

More generally, I am suggesting that Kant is one of those who think that

$$\text{Every finite set of Fs excludes at least one F,} \qquad (1)$$

though it contradicts the statement that there are only finitely many Fs, is nevertheless weaker than

$$\text{There is an infinite number of Fs.} \qquad (2)$$

Since (1) is weaker than (2), I think Kant thinks, the series of possible explorations can be more than finite without being infinite; and since the length of that series defines the size of the world, the world can escape being finite without being infinite.

I conjecture that Kant's reason for thinking that (1) is weaker than (2) is as follows. (1) is true if the Fs are the natural numbers, or the odd numbers, or the prime numbers, or the natural numbers > 7, or . . .; but if (1) = (2) then each of these sets has an infinite number of members, and so they all have the same number of members. Kant can be forgiven for assuming that there cannot be exactly as many prime numbers as odd numbers.

The assumption is of course a mistake, even if a forgivable one. We now know that by the only viable criterion of equal-numberedness there are as many primes as odd numbers. If this seems 'counter-intuitive', that is presumably because our intuitions about cardinality have been fed almost exclusively by our thinking about finite sets.[13] Kant, for one, carries finitistic assumptions over into his thinking about infinity. He assumes that an infinite number can count as an honest-to-God number only if it is 'determinate'; and, though 'determinate' is not explained, it seems fairly clear that for Kant a determinate number is one such that if you add one to it you get a different number. From this it follows that a 'determinate' number must be a finite one, i.e. that there cannot be an infinite number. I doubt if Kant sees this consequence of his assumptions. In a footnote he refers to 'a quantity (of given units) which is greater than any number' and says that this 'is the mathematical concept of the infinite' (A 432/B 460): this seems to imply that there cannot be an infinite number, but I am not sure how seriously to take this.

A significant importation of finitist thinking into a discussion of infinity occurs in a passage where Kant congratulates himself for not using a certain bad argument against the world's being infinitely old or large. The argument he didn't use is this:

A magnitude is infinite if a greater than itself, as determined by the multiplicity of given units which it contains, is not possible. Now no multiplicity is the greatest, since one or more units can always be added to it. Consequently an infinite given magnitude . . . is impossible. (A 430/B 458.)

The middle sentence is wrong, because adding 'one or more units' to an infinite number does not yield a higher number; but Kant voices no objection to this finitist intrusion. He quarrels only with the argument's first premiss:

The above concept is not adequate to what we mean by an infinite whole. It does not represent *how great* it is, and consequently is not the concept of a *maximum*. (A 430–2/B 458–60.)

This complaint that the proffered definition of 'infinite' does not 'represent *how great* it is' seems to mean that the definition doesn't define a determinate number – one which is just so large and no larger, this being thought of as the notion of a number n such that $n < (n + 1)$, and thus as the notion of a finite number. In short, having stayed silent on the argument's finitist error, Kant criticizes it on the basis of a finitist error of his own.

Kant is not the only philosopher to demand of infinite numbers a 'determinateness' which only finite numbers can have. Descartes says that 'in counting I cannot reach a highest of all numbers, and hence recognise that in enumeration there is something that exceeds my powers', from which he infers that 'a number is thinkable, that is higher than any that can ever be thought by me'.[14] Leibniz, too, shares Kant's nervousness about 'infinite number':

It is true that there is an infinity of things, i.e. that there are always more of them than can be assigned. But there is no infinite number, neither of line nor of other infinite quantity, if these are understood as veritable wholes . . . The true infinite exists, strictly speaking, only in the *absolute* which is anterior to all composition, and is not formed by the addition of parts.[15]

There is also a nice example of the same line of thought in Locke:

We have, it is true, a clear idea of division, as often as we think of it; but thereby we have no more a clear idea of infinite parts in matter, than we have a clear idea of an infinite number, by being able still to add new numbers to any assigned numbers we have: endless divisibility giving us no more a clear and distinct idea of actually infinite parts, than endless addibility (if I may so speak) gives us a clear and distinct idea of an actually infinite number: they both being only in a power still of increasing the number, be it already as great as it will. So that of what remains to be added (*wherein consists the infinity*) we have but an obscure, imperfect, and confused idea.[16]

I suspect that many other examples could be given, though these are all I have found so far. I should like to see a history of this matter. For example, it has been known since ancient times that every finite set of prime numbers excludes at least one prime number: I'd like to know when, and by whom, this was regarded as falling short of a proof that there is an infinite number of prime numbers.

IX. 'PRIOR TO ALL REGRESS'

How does this spurious 'weakening move' of Kant's relate to his official solution? Part of the answer is obvious. The official solution says that the world is not a thing in itself, which I am construing as an assertion of phenomenalism. This leads Kant to equate *the world's size* with *the length of the series of possible world-searches*, so that 'the world is not finitely large' becomes equated with 'Every finite world-search leaves some world unsearched'; and this has nested within it the supposedly weaker-than-infinity statement that no finite amount of world includes all the world there is.

This link between the official solution and the 'weakening move' is an accidental one. The thought that 'Every finite world-stretch excludes some world' conflicts with 'The world is finite' without entailing 'The world is infinite' – this thought might have occurred to Kant in just that form, without his being led to 'Every finite world-stretch excludes some world' through its being nested within the phenomenalistic 'Every finite world-search leaves some world unsearched'.

Still, the official solution connects with the 'weakening move' in another way as well, for Kant has certain formulations which can express *both* the thesis that the world's extent is not infinite *and* the pseudo-thesis that the world is not a thing existing in itself. Here is a crucial passage:

[We must not regard the world] as a thing given in and by itself, prior to all regress. We must therefore say that the number of parts in a given appearance is in itself neither finite nor infinite. For an appearance is not something existing in itself, and its parts are first given in and through the regress of the decomposing synthesis, a regress which is never given in absolute completeness, either as finite or as infinite. (A 505/B 533.)

Kant is there concerned with infinite divisibility (the 'decomposing synthesis') but the passage also bears directly on my present topic. When Kant says that the world isn't given 'prior to all regress', he means: statements about possible experience are not mere consequences of independently intelligible facts about the world, but rather give to statements about the world all the content we can understand them as having. But combined with that thought there is also the following different one. The statement that no finite set exhausts the members of a given series mustn't be thought of as a consequence of the series' having a determinate infinite number of members; rather, the statement that no finite set exhausts the series *is* the strongest statement we can make abut the size of the series.

I think that both elements are present in this passage:

The cosmic series can . . . be neither greater nor smaller than the possible empirical regress upon which alone its concept rests. And since this regress can yield neither a determinate infinite nor a determinate finite . . ., it is evident that the magnitude of the world can be taken neither as finite nor as infinite. The regress, through which it is represented, allows of neither alternative. (A 518n./B 546 n.)

I suggest that Kant is here playing with two thoughts at once, both having the form 'Our only concept of *x* is our concept of our approach to *x*'. One thought is that our only concept of the world is that of our actual and possible experiences of the world, while the other thought is that our only concept of a non-finite series is the concept of a series some of which always lies ahead of us.

The most striking example of all is in the following passage:

> Of this empirical regress the most that we can ever know is that from every given member of the series of conditions we have always still to advance empirically to a higher and more remote member. The magnitude of the whole of appearances is not thereby determined in any absolute manner; and we cannot therefore say that this regress proceeds to infinity. In doing so we should be anticipating members which the regress has not yet reached, representing their number as so great that no empirical synthesis could attain thereto, and so should be determining the magnitude of the world (although only negatively) prior to the regress – which is impossible. (A 519/ B 547.)

The mistake of 'representing the world's magnitude as so great that no empirical synthesis could attain thereto' is really two mistakes: thinking that questions about the world's magnitude concern something more than actual and possible experience; and thinking that we could say something about the world's magnitude stronger than that any finite stretch of world excludes some world. These are indeed both mistakes. But they have no direct and straightforward connection with one another. Kant thinks that they have, and indeed tends to identify them with one another, only because he has been misled by the protean phrase 'prior to the regress' and others like it.

X. THE FUTURIZING MOVE

There is a second major element in Kant's purportedly problem-solving material – an element which more directly involves his phenomenalism. In exhibiting it, I take my stand on Sections IV–VI above: the infinities Kant finds troublesome are all and only those which don't lie in the future.

Let us look back at the phenomenalizing move as applied to the size of the world. The statement that the world is more than finitely large is equated with the statement that however much world I explore there will always be more world to be explored. Kant thinks that this stops short of saying that the world is infinitely large; but even if he didn't, and still said that the world is infinite in extent, he would still be left with a problem-solving remark to make – namely that the relevant infinity is now projected into the future. This draws the sting from 'The world is infinitely large', because the latter is now equated with a statement about a possible *future* infinite series. According to the phenomenalist analysis, my thought of the world as being infinitely large is my thought of embarking upon a never-ending non-repetitive series of world-explorations – such a series being, though infinite, conceptually harmless because lying wholly in a possible future.

So much for the infinity of the world's present extent; and the same pattern of problem-solution applies even to the most completed-seeming infinity of all, namely the infinite series of past events. For if we take the statement that this series is infinite, and ask about it the question that causes all the trouble – namely 'What does this statement mean in terms of what I could discover for myself?' – we find that the answer projects even the series of past events into the future, or into a possible future. That is, the idea of the series of past events is the idea of what I should discover if (in the future) I pursued my researches deeper and deeper into the past. Thus Kant:

> The real things of past time . . . are objects for me and real in past time only in so far as I represent to myself (either by the light of history or by the guiding clues of causes and effects) that a regressive series of possible perceptions in accordance with empirical laws, in a word, that the course of the world, conducts us to a past time-series as condition of the present time – a series which, however, can be represented as actual not in itself but only in the connection of a possible experience. Accordingly, all events which have taken place in the immense periods that have preceded my own existence mean really nothing but the possibility of extending the chain of experience from the present perception back to the conditions which determine this perception in respect of time. (A 495/B 523.)

In that passage, I think, Kant is making problem-solving remarks without properly grasping just what it is in them which solves the problem. This is hardly surprising since, as we have seen, he does not even succeed in explaining clearly what the problem is. There are other passages, too, where Kant subjects the troublesome infinity-statements to operations which project the relevant infinities into the future, and where he then relaxes. Here are two more examples (emphases mine):

> If I represent to myself all existing objects of the senses in all time and in all places, I do not set them in space and time [as being there] prior to experience. This representation is nothing but the thought of a possible experience in its absolute completeness. Since the objects are nothing but mere representations, only in such a possible experience are they given. To say that they exist prior to all my experience is only to assert that they are *to be met with* if, starting from perception, I *advance* to that part of experience to which they belong. (A 495–6/B 523–4.)
> To call an appearance a real thing prior to our perceiving it, either means that in the *advance* of experience we must meet with such a perception, or it means nothing at all. (A 493/B 521.)

It is true that Kant does not explicitly set up his problem as one about non-future infinities as such; and it is also true that many of his remarks imply that there can be no legitimate use of the concept of infinity – indeed that there really isn't any such concept. But I still contend that a good part of his sense of having solved his problem is due to the fact that he is troubled about non-future infinities as such and the fact that he can see how, through phenomenalist analyses, to throw these infinities into the future.

[1] P. F. Strawson, *The Bounds of Sense* (London: Methuen, 1966), p. 176.

[2] E. Caird, *The Philosophy of Kant* (Glasgow: Maclehose, 1877), p. 567.

[3] N. Kemp Smith, *A Commentary to Kant's Critique of Pure Reason* (London: Macmillan, 2nd ed. 1923), p. 484.

[4] A. C. Ewing, *Short Commentary on Kant's Critique of Pure Reason*, (London: Methuen, 1938), 211–12.

[5] T. D. Weldon, *Kant's Critique of Pure Reason*, (Oxford: Clarendon Press, 2nd ed. 1958), 205–6.

[6] 'The impossibility of an actual infinite rests in the last resort on the world being created by God and, as God-created, being a world that is finite throughout,' G. Martin, *Kant's Metaphysics and Theory of Science* (Manchester University Press, 1955), 50.

[7] José A. Benardete, *Infinity* (Oxford: Clarendon Press, 1964), 121–2.

[8] Strawson, *op. cit.*, p. 177. The best discussion of the First Antinomy, apart from Strawson's, is to be found in C. D. Broad, 'Kant's Mathematical Antinomies', *Proceedings of the Aristotelian Society*, **55**, (1954–5).

[9] Weldon, 206.

[10] Kemp Smith, 484.

[11] Fred Dretske, 'Counting to Infinity', *Analysis* XXV (1964–5).

[12] Reported to me by Elizabeth Anscombe.

[13] Cf. Bertrand Russell, 'Mathematical Infinity', *Mind* N. S. LXVII (1958).

[14] Descartes, in Haldane and Ross (eds.), *Philosophical Works of Descartes* (Cambridge University Press, corrected ed. 1934), II. 37–8.

[15] Leibniz, *New Essays* II. xvii. 1.

[16] Locke, *Essay* II. xxix. 16.

NOTES ON THE CONTRIBUTORS

CHARLES PARSONS is Professor of Philosophy at Columbia University, New York.

JAMES HOPKINS is Lecturer in Philosophy at King's College, London.

DIETER HENRICH is Professor of Philosophy at the University of Heidelberg.

SIR PETER STRAWSON is Waynflete Professor of Metaphysical Philosophy in the University of Oxford, and a Fellow of Magdalen College, Oxford.

LAUCHLAN CHIPMAN is Professor of Philosophy at the University of Wollongong, New South Wales.

BARRY STROUD is Professor of Philosophy at the University of California at Berkeley.

H. E. MATTHEWS is Senior Lecturer in the Department of Logic at the University of Aberdeen.

W. H. WALSH retired in 1979 as Professor of Logic and Metaphysics at the University of Edinburgh, and as Vice-Principal of the University. He now lives in Oxford.

JONATHAN BENNETT is Professor of Philosophy at Syracuse University, New York.

BIBLIOGRAPHY

(not including material in this volume)

1. INTRODUCTORY BOOKS

Kemp, J., *The Philosophy of Kant* (Oxford University Press, 1968).
Körner, S., *Kant* (Harmondsworth: Penguin, 1955).

2. SOME GENERAL BOOKS

Bennett, J., *Kant's Analytic* (Cambridge University Press, 1966).
Bennett, J., *Kant's Dialectic* (Cambridge University Press, 1974).
Bird, G., *Kant's Theory of Knowledge* (London: Routledge, 1962).
Broad, C. D., *Kant: an Introduction* (Cambridge University Press, 1978).
Ewing, A. C., *A Short Commentary on Kant's Critique of Pure Reason* (London: Methuen, 2nd ed. 1950).
Smith, N. Kemp, *A Commentary to Kant's 'Critique of Pure Reason'* (London: Macmillan, 2nd ed. 1923).
Strawson, P. F., *The Bounds of Sense* (London: Methuen, 1966).
Walker, R. C. S., *Kant* (London: Routledge, 1978).
Walsh, W. H., *Kant's Criticism of Metaphysics* (Edinburgh University Press, 1975).
Wilkerson, T. E., *Kant's Critique of Pure Reason* (Oxford: Clarendon Press, 1976).
Wolff, R. P., *Kant's Theory of Mental Activity* (Harvard University Press, 1963).

3. COLLECTIONS OF ESSAYS

Beck, L.W., *Kant Studies Today* (Lā Salle: Open Court, 1969).
Gram, M. S., *Kant: Disputed Questions* (Chicago: Quadrangle Books, 1967).
Penelhum, T., and MacIntosh, J. J., *The First Critique* (Belmont: Wadsworth, 1969).
Wolff, R. P., *Kant* (New York: Doubleday Anchor, 1967).

4. Particular Topics

(In this section full publication details are not given for books already listed above)

(a) Arithmetic

Barker, S. F., *Philosophy of Mathematics* (Englewood Cliffs: Prentice-Hall, 1964), pp. 72–7.

Martin, G., *Kant's Metaphysics and Theory of Science* (Manchester University Press, 1955), ch. 1.

Hintikka, J., 'Kant on the Mathematical Method', in Beck (ed.), *Kant Studies Today*.

Hintikka, J., 'On Kant's Notion of Intuition (Anschauung)', in Penelhum and MacIntosh (eds), *The First Critique*.

Kitcher, P., 'Kant and Mathematics', *Philosophical Review* **84** (1975).

Thompson, M., 'Singular Terms and Intuitions in Kant', *Review of Metaphysics* **26** (1972–3).

Craig, E., 'An Approach to the Problem of Necessary Truth', in S. Blackburn (ed.), *Meaning, Reference and Necessity* (Cambridge University Press, 1975).

(b) Geometry (see also under *(a)*, above)

Barker, S. F., *Philosophy of Mathematics* (Englewood Cliffs: Prentice-Hall, 1964), chs 2 and 3.

Bennett, J., *Kant's Analytic*, ch. 2

Strawson, P. F., *The Bounds of Sense*, part 5.

Reichenbach, H., *The Philosophy of Space and Time* (New York: Dover, 1957), ch. 1.

Lucas, J. R., 'Euclides ab omni Naevo vindicatus', *British Journal for the Philosophy of Science* **20** (1969).

Craig, E., 'Phenomenal Geometry', *British Journal for the Philosophy of Science* **20** (1969).

Nerlich, G., *The Shape of Space* (Cambridge University Press, 1976), chs 3 and 4.

(c) On whether Experience must be Spatial

Quinton, A., *The Nature of Things* (London: Routledge, 1973), ch. 1.

Strawson, P. F., *Individuals* (London: Methuen, 1959), chs 1, 2 and 4.

Strawson, P. F., *The Bounds of Sense*, pp. 125–32.

Bennett, J., *Kant's Analytic*, ch. 3.

Evans, G., 'Things Without the Mind', in Zak van Straaten (ed.), *Philosophical Subjects: Essays Presented to P. F. Strawson* (Oxford: Clarendon Press, 1980).

(d) Transcendental Deduction of the Categories

Mackie, J. L., *The Cement of the Universe* (Oxford: Clarendon Press, 1974), ch. 4.

Ewing, A. C., *A Short Commentary on Kant's Critique of Pure Reason*, ch. 3.

Paton, H. J., 'The Key to Kant's Deduction of the Categories', *Mind* N. S. **40** (1931); also in Gram (ed.), *Kant: Disputed Questions.*

Bennett, J., *Kant's Analytic*, chs 8 and 9.

Strawson, P. F., *The Bounds of Sense,* part 2, ch. 2.

Wolff, R. P., *Kant's Theory of Mental Activity,* pp. 78–202; partly reprinted as 'A Reconstruction of the Argument of the Subjective Deduction' in Wolff (ed.), *Kant.*

Henrich, D., *Identität und Objektivität* (Heidelberg: Carl Winter Universitätsverlag, 1976).

(e) Schematism

Warnock, G. J., 'Concepts and Schematism', *Analysis* **9** (1948–9).

Walsh, W. H., 'Schematism', *Kant-Studien* **49** (1957–8); also in Wolff (ed.), *Kant.*

Wilkerson, T. E., *Kant's Critique of Pure Reason*, ch. 4, §8.

Wolff, R. P., *Kant's Theory of Mental Activity*, pp. 206–23.

(f) Transcendental Arguments

Walsh, W. H., 'Philosophy and Psychology in Kant's Critique', *Kant-Studien* **57** (1966).

Harrison, R., *On What There Must Be* (Oxford: Clarendon Press, 1974), chs 1 and 2.

Wilkerson, T. E., *Kant's Critique of Pure Reason*, ch. 10.

Strawson, P. F., *Individuals* (London: Methuen, 1959), esp. pp. 34–5.

Strawson, P. F., *The Bounds of Sense,* esp. part 1, §§1–2; part 2, ch. 2, §§1 and 4.

Körner, S., 'The Impossibility of Transcendental Deductions', *Monist* **51** (1967); also in Beck (ed.), *Kant Studies Today.*

Hacker, P. M. S., 'Are Transcendental Arguments a Version of Verificationism?' *American Philosophical Quarterly* **9** (1972).

Walker, R. C. S., *Kant,* ch. 2.

(g) Causal Law

Mackie, J. L., *The Cement of the Universe* (Oxford: Clarendon Press, 1974), ch. 4.

Strawson, P. F., *The Bounds of Sense,* part 2, ch. 3.

Bennett, J., *Kant's Analytic,* chs 11 and 15.

Bennett, J.,'Strawson on Kant', *Philosophical Review* **77** (1968), pp. 345–6.

Walsh, W. H., 'Kant on the Perception of Time', *Monist* **51** (1967); also in Beck (ed.), *Kant Studies Today;* also in Penelhum and MacIntosh (eds), *The First Critique.*

Ayer, A. J., *The Foundations of Empirical Knowledge* (London: Macmillan, 1940), §§17–20.

(h) Transcendental Idealism

Strawson, P. F., *The Bounds of Sense,* parts 1 and 4.

Bennett, J., *Kant's Analytic,* esp. §§7, 8, 18, 32, 52.

Bennett, J., *Kant's Dialectic,* esp. §§16–18.

Chipman, L., 'Things in Themselves', *Philosophy and Phenomenological Research* **33** (1972–3).

Wilkerson, T. E., *Kant's Critique of Pure Reason,* ch. 9.

Bird, G., *Kant's Theory of Knowledge,* chs 1–5.

Sellars, W., *Science and Metaphysics* (London: Routledge, 1968), ch. 2.

Walker, R. C. S., *Kant,* chs 9 and 12.

Adickes, E., *Kant und das Ding an Sich* (Berlin: Pan, 1924), esp. ch. 1.

(i) The Self

Strawson, P. F., *The Bounds of Sense,* part 3, ch. 2.

Broad, C. D., *Kant: an Introduction,* pp. 234–63.

Bennett, J., *Kant's Dialectic,* chs 3–6.

Walsh, W. H., *Kant's Criticism of Metaphysics,* §§31–2.

Wilkerson, T. E., 'Kant on Self-Consciousness', *Philosophical Quarterly* **30** (1980).

(j) The First Two Antinomies

Broad, C. D., *Kant: an Introduction,* pp. 210–33; essentially a reprint of his article 'Kant's Mathematical Antinomies', *Proceedings of the Aristotelian Society* **55** (1954–5).

Strawson, P. F., *The Bounds of Sense,* part 3, ch. 3.

Bennett, J., *Kant's Dialectic,* chs 7–9.

Swinburne, R., *Space and Time* (London: Macmillan, 1968), chs 14 and 15.

Parsons, C. D., 'Infinity and Kant's Conception of the "Possibility of Experience" ', *Philosophical Review* **73** (1964); also in Wolff (ed.), *Kant.*

Russell, B. A. W., *The Principles of Mathematics* (London: Allen & Unwin, 2nd ed. 1937), ch. 52.

(k) Freedom

Campbell, C. A., 'Is "Free Will" a Pseudo-Problem?', *Mind* N. S. **60** (1951).

Smart, J. J. C., 'Free Will, Praise and Blame', *Mind* N. S. **70** (1961).

Strawson, P. F., 'Freedom and Resentment', *Proceedings of the British Academy* 48 (1962); also in Strawson (ed.), *Studies in the Philosophy of Thought and Action* (Oxford University Press, 1968); also in Strawson, *Freedom and Resentment and Other Essays* (London: Methuen, 1974).

Bennett, J., *Kant's Dialectic,* ch. 10.

Beck, L. W., *A Commentary on Kant's Critique of Practical Reason* (University of Chicago Press, 1960), ch. 11.

Körner, S., 'Kant's Conception of Freedom', *Proceedings of the British Academy* **53** (1967).

(l) God: the Ontological Argument

Ryle, G., 'Systematically Misleading Expressions', *Proceedings of the Aristotelian Society* **32** (1931–2); also in A. Flew (ed.), *Logic and Language,* First Series (Oxford: Blackwell, 1951); also in Ryle, *Collected Essays* (London: Hutchinson, 1971), vol. II.

Pears, D., and Thomson, J., 'Is Existence a Predicate?' in Strawson (ed.), *Philosophical Logic* (Oxford University Press, 1967).

Kenny, A. J. P., *Descartes* (New York: Random House, 1968), ch. 7.

Barnes, J., *The Ontological Argument* (London: Macmillan, 1972), chs 3 and 4.

Alston, W. P., 'The Ontological Argument Revisited', *Philosophical Review* **69** (1960); also in A. Plantinga (ed.), *The Ontological Argument* (Garden City: Doubleday, 1965); also in W. Doney (ed.), *Descartes* (New York: Doubleday Anchor, 1967).

Strawson, P. F., 'Is Existence Never a Predicate?', *Critica* **1** (1967); also in Strawson, *Freedom and Resentment and Other Essays* (London: Methuen, 1974).

Mackie, J. L., 'The Riddle of Existence', *Proceedings of the Aristotelian Society* Supplementary Vol. **50** (1976).

(m) God: the Cosmological Argument

Hume, D., *Dialogues concerning Natural Religion* (London, 1779), §IX.

Plantinga, A., *God and Other Minds* (Ithaca: Cornell University Press, 1967), ch. 1.

Martin, C. B., *Religious Belief* (Ithaca: Cornell University Press, 1959), ch. 9.

Bennett, J., *Kant's Dialectic*, §§75–80.
Strawson, P. F., *The Bounds of Sense*, part 3, ch. 4.
Swinburne, R., *The Existence of God* (Oxford: Clarendon Press, 1979), ch. 7.

INDEX OF NAMES

(not including authors mentioned only in the Bibliography)